Scienatics

The science of science

Naiyf Saud

Agio
PUBLISHING HOUSE

Agio
PUBLISHING HOUSE

BC Canada
www.agiopublishing.com

Scienatics, The Science of Science
ISBN 978-1-927755-68-6 (paperback)
ISBN 978-1-927755-69-3 (ebook)

Cataloguing information available from
Library and Archives Canada.
Printed on acid-free paper.
Agio Publishing House is a socially-responsible enterprise, measuring success on a triple-bottom-line basis.

10 9 8 7 6 5 4 3 2 1

Contents

Preface

The nature of science is an elusive concept, so any attempt to define science did not succeed in stipulating the extension of that concept such that all of its members are included. The common practice of defining science by its activities is not appropriate, since the activities involving different sciences are different and hence the members, which constitute the class of reference of science, are distributed in several extensions that appear to have different nature while all of them refer to the same concept. Also this type of definitions is, in fact, a definition of the practice and intellectual activities of science rather than an exact definition of science. On the other hand, defining science as a collection of principles related to a particular subject produces a huge confusion since, according to this definition, any collection of concepts, without any prescribed plausible justification, can be regarded as a science. These aspects have the signature of the lack of understanding of the true nature of science which was produced by our unconscious ignorance of key factors that have important clues about the nature of this concept. For instance, if we try to study the relation between the pressure of a gas and its volume, we have to pay attention to its temperature. Otherwise, the variable temperature would produce a chaotic pattern for that relation. This character of science arises from its interference with other concepts that appear to have no connection with it. These concepts increase our appreciation of the nature of existence and their relation to science and reality. This book is an attempt to develop a theory that treats science more precisely and hence reveals its genuine aspects and constitutional structure. This theory will enhance our instinct to distinguish between science, pseudoscience and fantasy and explain the scienatic phenomena that we encounter every time we deal with definitions, equations and numbers. In the first chapter, we discuss the nature of existence since it is a central idea for our understanding of science as one of the five features of any concept. The seven types of concepts and their three aspects for the orthodox observers are also investigated in this chapter. In chapter two, we are investigating the Adamic sciences, their constituents and the constitutional structure of science. The essential and accidental contents of science are also investigated in this chapter. We end this book by discussing the distinction between science, pseudoscience and fantasy and the relations between mind, science and logic which constitute the intellectual system.

Naiyf S. Alsaud
March 2018

1 Introduction

1.1 The nature of existence

One of the most speculative and perplexing inquiries is the quest to reveal the essences of entities. Questions like: What are the ultimate natures of light, sound, scents, tastes and tangibles or the absolute essences of brightness, colors, voice, smell, flavors and warmness? What are the inherent natures of time and space? What are the essences of virtues and vices? have caught the attentions of humankind fancy along the development of their civilization. However, the products of human intellectual activities have revealed counter-intuitive answers to those questions. The intellectual heritage accumulated in many disciplines provides a strong evidence for these unexpected results. What we see as a light depends on the sensitivity of our photoreceptor cells in the retina which was developed as a ramification of the peak radiation of the sun. So it is not a coincidence that we see in the electromagnetic waves that have wavelengths in the visible light. This sensitivity could be developed, in principle, for any wavelength range of the electromagnetic radiation if the environment in which the organism has evolved was dominated by radiations in that range. This well-established result indicates that what we call light and darkness does, in fact, depend on the structure of our optical detectors and brains. We could perform a thought experiment to appreciate the relativistic character of this phenomenon. Suppose that there were two organisms - one of them was evolved in an environment that acquired him the sensitivity of the electromagnetic waves in the range $\sim 10 - 380$ nm, which lies in the ultraviolet radiation, and the other evolved in Earth, so its photoreceptor cells are sensitive to the spectrum in the range $\sim 390 - 700$ nm. Further suppose that they were setting in a room which has two light bulbs, one emits ultraviolet radiation and the other visible light. Now if we turn on the former light bulb, the organism whose optical detectors are sensitive to the ultraviolet light would see everything in the room and perceive the aspects of lightness after darkness while the situation remains unchanged for the terrestrial organism. The situation would be exactly the opposite if we instead have turned on the latter light bulb which radiates in the visible range. On the other hand, the acuteness of their sensitivity to light will govern their judgment about the brightness of their respective bulbs. Hence if the acuteness of the sensitivity to the visible light of the terrestrial was more than that of the other organism, he would see his bulb brighter than the brightness of the ultraviolet bulb that was experienced by the alien, even if both of them have the same brightness. The identities of the colors which they describe are determined by the structures of their brains which were shaped by the different environments in which they have evolved. So there are no absolute natures of the primary colors (red, yellow and blue) and their mixtures but rather relativistic aspects which were created by their inherently different brains. In the case that there were sources of sounds, scents and tastes in that room, the essential natures of them depend on the structures of the creatures senses and the interpretation of their brains which is a reflection of the conditions in which they have developed. The differences in the essences of these observables between the two organisms could, hypothetically, be enormous such that some or all of the observables for one organism perform different functionalities for the other. For instance, the longitudinal waves that carry the vibration through a medium as a sound for one organism could be used as a mean of vision (the standard role of light) for another and the same proposition is true for the other observables. Moreover, their descriptions and feelings of the temperature of the room depend on their respective normal body temperatures. So if the temperature of the room was 15° C and the alien creature and the terrestrial organism had a normal body temperature of 10° C and 37° C respectively, the alien would describe the room as a warm place while it would be a cold place for the terrestrial. This demonstrates that what we call a nice cool summer evening would be a harsh condition for another organism and vice versa.

The essence of time has puzzled the ordinary people as well as the intellectual communities through ages. The impenetrable comprehension of time emerges from its duality of existence and intangible character. Its characteristic of being measured produces the feeling of its existence and its intangible aspect conceals its nature. We use days, hours, minutes and seconds to measure time which are, traditionally, defined by letting the day be the period of a complete rotation of Earth about its axis and define the hour as one part of 24 parts of that period and define minute as one part of 60 parts of an hour and so on or, recently and more accurately, by using the transition of the electrons between the two hyperfine levels of the ground state of the cesium-133 atom, we defined the second as 9192631770 periods of that transition and define the minute as 60 seconds and so on. However, in both definitions or in any possible inherently similar definitions, we are, actually, measuring "changes" which is a measurand that is ordered by nature, so there are changes with low levels of intensity and changes with high levels of intensity and infinitely many levels of changes between them. The apparent proper motion of a star as a result of a complete spin of Earth is a change by 360° and the transition of an electron between two hyperfine levels is a change in its position. Moreover, the intensity of the former is much greater than the intensity of the latter which characterizes change as a measurand. However, the possibility of estimating the entirety of a given level of intensity by its parts is a property of any measurand. To estimate any length, we use standard units of length (e.g., meter, foot or yard) which themselves are lengths. Also, to measure mass we utilize standard units of mass (gram, pound, ounce, etc.) as standard units which themselves constitute masses. This reasoning could be carried to any measurand which leads, correctly, to the conclusion that the standard units which we use to measure change should themselves constitute kinds of changes!! So if we would like to measure the speed of an object, which is a change in position, we could use the step of a person walking on the pavement, which is also a change in position, and say: for each step of that person the object moves 5 km. Also we could compare the intensities of the movements of that object and other objects if we use a unique standard change as a unit (e.g., the step of that person) to measure their movements and, hence time in its essential nature is change and its units (e.g., hour, minute and second) are standard units of change. However, for the historical development within the orthodox line of thought, we coined the concept "time" to indicate that standard change which we use to measure and compare changes. If we consider any dynamical phenomenon such as growth, heating or even painting, we find that its essential nature is movement. Growth occurs when the living cells undergo subsequent changes which are results of the movements of their ultimate constituents. The increase in the temperature of a body by heating is a reflection of the increase of the average translational velocity of its molecules which is in its essence a movement. Painting a picture involves the transformation of the paints from the palette to the portrait and producing artistic patterns by distributing them rationally in the portrait using the brush and so all of the subsequent changes to completely draw a picture are, in their essences, movements. This sample of dynamical phenomena, which certainly involve changes, can be expanded to include more and more types of change and hence inductively supports the conclusion that any type of change is essentially a movement.

The deduced universal upper limit of movement as a ramification of the constancy of the speed of light, which is confirmed experimentally by Michelson-Morley, should have effects on any type of change including time since the essential nature of the latter is a movement. To deduce the nature of the influence of this phenomenon on the movement and hence on any change, we begin with Lorentz transformations which are the logical consequences of the constancy of the speed of light, the relativistic principle (the laws of physics are the same in all inertial frames of reference) and the isotropy and homogeneity of space. We state these transformations below: (They can be easily deduced based on the previously mentioned three assumptions.)

$$t' = \frac{t - \frac{vx}{c^2}}{\sqrt{1 - \frac{v^2}{c^2}}}, \quad x' = \frac{x - vt}{\sqrt{1 - \frac{v^2}{c^2}}}, \quad y' = y, \quad z' = z \qquad (1.1)$$

where c is the speed of light and x, y, z and t are the respective space and time coordinates of the subsequent events in space-time for the stationary frame of reference S, and x', y', z' and t' are the corresponding coordinates of the same sequence of events for a frame of reference S′ which have, with respect to S, a relative velocity v along the x direction. (We have assumed that $t = t'$ when $x = x'$ which does not produce any loss of generality.) Now by differentiating these equations with respect to t we get:

$$\frac{dt'}{dt}=\frac{1-\frac{v\frac{dx}{dt}}{c^2}}{\sqrt{1-\frac{v^2}{c^2}}},\quad \frac{dx'}{dt}=\frac{\frac{dx}{dt}-v}{\sqrt{1-\frac{v^2}{c^2}}} \tag{1.2}$$

and

$$\frac{dy'}{dt}=\frac{dy}{dt},\quad \frac{dz'}{dt}=\frac{dz}{dt} \tag{1.3}$$

so by dividing the latter derivatives of each spatial coordinate of S$'$ by its derivative of time and using the definition of velocity we have:

$$u'_x=\frac{u_x-v}{1-\frac{vu_x}{c^2}},\quad u'_y=u_y\frac{\sqrt{1-\frac{v^2}{c^2}}}{1-\frac{vu_x}{c^2}},\quad u'_z=u_z\frac{\sqrt{1-\frac{v^2}{c^2}}}{1-\frac{vu_x}{c^2}} \tag{1.4}$$

It is obvious from these equations that any movement in the direction of the relative velocity of the two frames (along the x-axis) or in the directions perpendicular to it (along the y or z axes), will be continuously reduced to lower intensities as the relative velocity between them reaches the speed of light. In fact this proposition is true for the movement in any direction, since, in the three dimensional space, this movement can be analyzed into three components along the Cartesian-axes and the orientation of these coordinates can be set arbitrarily. So if we choose the orientation of the coordinates to coincide with those used in our frame of references, each of these components would be reduced as we have shown and hence their resultant (the movement) will be reduced as the relativistic velocity of the frames approaches the speed of light. This effect emphasizes that, relative to a stationary frame of reference, any movement and hence any change will be diminished to lower and lower intensities as the velocity of its frame of reference increases to more and more values. However, this phenomenon will not be pronounced until the relativistic velocity becomes sufficiently close to the upper intensity of movement (the speed of light). Hence, if we imagine an astronaut traveling in a spacecraft with a speed very close to the speed of light such that the effects of the previous phenomenon are significant, the intensity of any change in the spacecraft will be diminished by the same rate (relative to an observer in Earth). The movements of the oscillating elements in his clocks on which the seconds and hence minutes and hours are based will be reduced, so time (the standard change) will dilate. This rate of reduction will be experienced by the movements of the nervous signals that carry the information to and from his brain, along with the movements of the atoms and the electrons binding them in molecules which constitute the ultimate microscopic bases that produce the biochemical reactions which manifest themselves as his growth and metabolism. The same arguments are applied to his movements or any other type of movements (changes) in his spacecraft. As a result, any change in the spacecraft (time, growth, reaction, etc.) will appear to be slower than its counterpart in Earth by the same rate. This line of reasoning can be carried to the observer on Earth relative to the astronauts since the former has the same relativistic velocity with respect to the latter. However, the reasoning of the observer on Earth is genuine and that of the astronaut is an illusion which is a reflection of the first. This is true because, before the astronaut left the Earth, the physical situations of the observer and the astronaut universally (relative to any thing in the universe) were the same, but after the astronaut has left the Earth, the universal physical situation of the astronaut diverged from its initial state while the observer's universal physical situation remained unchanged in the same state and so if we claimed that the astronaut's reasoning was genuine, the observer's initial condition should reach a new condition without any change in its universal initial physical situation which is imposable. (We always get nothing from nothing.) However, the astronaut's universal physical situation has changed to a new situation and this "genuine new change" leads to a universally new genuine condition, so the astronaut's reasoning is an illusion which is a reflection of the genuine reasoning of the observer on Earth. Therefore, any deduction based on the illusive argument would lead to false conclusions and any deduction based on the genuine argument would lead to real conclusions. As a consequence of that, the astronaut would reach the real conditions if he assumed that his observations are illusions and in reality the Earth is stationary while he is moving with a relativistic velocity with respect to it and performed his reasoning and calculations using the relativistic formulas on this basis. On the other hand, the observer on Earth can assume the reality of his observations and deduce his relativistic real conclusions based on that assumption. Therefore, the criterion of distinction between the illusive and real conditions is the nature of the universal difference between the initial and final physical states of the frame of references. As we have mentioned the upper bound existence of the intensity of movements leads to the reduction of the intensity of any change and hence stretches the intervals of time in a phenomenon called "time dilation". We could deduce the latter effect mathematically as follows:

$$\Delta t' = \int_{t'_1}^{t'_2} dt' = \int_{t_1}^{t_2} \frac{dt}{\sqrt{1-\frac{v^2}{c^2}}} = \frac{\Delta t}{\sqrt{1-\frac{v^2}{c^2}}} \tag{1.5}$$

where $\Delta t'$ is the time internal in the stationary frame of reference and Δt is the corresponding interval with respect to the moving frame. This effect entails logically the well know phenomenon of "twin paradox". According to which two twins would grow and respond to stimulation with different rates if each one of them lived in a different inertial frame of reference with non-zero relativistic velocity between them. However, this phenomenon, as mentioned, will not be pronounced except if their relative velocity contributes a significant portion of the speed of light. Now if we assumed that this effect of the relativistic velocity on time appreciation was produced by the natural structure of one or both of the observers, the two observers would acquire different senses of time, even if they were on the same frame of reference. These arguments strongly support the conclusion that even the essential nature of time depends on the natural structure of the observer.

In mathematics any set of elements (called points) for which the mathematical concepts of continuity, connectedness and convergence can be defined is called a "topological space". This most abstract notion of space constitutes the ultimate resemblance of the Euclidean space and so these aspects are the least properties that a set of points (which could be any type of elements) should have to be described mathematically as a space. The topological space is called a metric space if the distances between its points can be defined and, in the case that this metric processes the properties of the Euclidean metric, it is considered a normed vector space. Finally it is called an inner product space if it, above that, has the notions of the vectors lengths and the angles between them which is a ramification of having an inner product. It is obvious that the concept of space was originally created to describe the set of points which constitute the Euclidean space, so the latter has the definitional aspects of any space and hence it is a normed vector space. We will consider the possible normed vector spaces that consist of "geometrical points" including the Euclidean space. The distance from an arbitrary point to another, in any geometrical space, can be described quantitatively by constructing a coordinate system based on any arbitrary path connecting them and so there are infinitely many possible coordinate systems that could be utilized to accomplish this task. The metric tensor is a mathematical concept that describes the relation between the coordinate system and the distance. So, for a given geometrical space, the elements of the metric tensor will change as a consequence of changing the coordinate system while the distance which is an intrinsic geometrical property will remain unchanged. If we were using the Cartesian coordinate system, the metric tensor of the Euclidean space would have the form:

$$g_{ij} = \begin{bmatrix} 1 & 0 & 0 \\ 0 & 1 & 0 \\ 0 & 0 & 1 \end{bmatrix} \tag{1.6}$$

however, based on the related definitions, it can be shown that the square of the distance in any coordinate system is given by:

$$ds^2 = g_{ij} dx^i dx^j \tag{1.7}$$

and hence, in the Cartesian coordinates, we would have:

$$\begin{aligned} ds^2 &= \begin{bmatrix} 1 & 0 & 0 \\ 0 & 1 & 0 \\ 0 & 0 & 1 \end{bmatrix} \times \begin{bmatrix} dx \\ dy \\ dz \end{bmatrix} \times \begin{bmatrix} dx \\ dy \\ dz \end{bmatrix} \\ &= dx^2 + dy^2 + dz^2 \end{aligned} \tag{1.8}$$

Now since the distance is invariant for a given geometrical space, we would have for any two coordinate systems:

$$g'_{\alpha\beta} dx^\alpha dx^\beta = ds^2 = g_{ij} dx^i dx^j \Rightarrow g'_{\alpha\beta} = \frac{dx^i}{dx^\alpha} \frac{dx^j}{dx^\beta} g_{ij} \tag{1.9}$$

This formula can be used to calculate the metric tensor for a geometrical space in a coordinate system given the tensor in another coordinate system and the relations between them. We know by definition that the relations between the spherical and Cartesian coordinates are:

$$x = r \sin\theta \cos\phi, \quad y = r \sin\theta \sin\phi, \quad z = r \cos\theta \tag{1.10}$$

so using these relations and the metric tensor in Cartesian coordinates along with the previous formula, the fundamental tensor in the spherical coordinates would be:

$$g'_{\alpha\beta} = \begin{bmatrix} 1 & 0 & 0 \\ 0 & r^2 & 0 \\ 0 & 0 & r^2 \sin^2\theta \end{bmatrix} \tag{1.11}$$

therefore, based on this tensor, the distance in the Euclidean space described by the spherical coordinates will be given by:

$$ds^2 = dr^2 + r^2 d\theta^2 + r^2 \sin^2\theta d\phi^2 \tag{1.12}$$

The last equation constitutes the metric in a continuous flat three-dimensional space which has a number of aspects that are consequences of its fundamental properties and hence there are hypothetically infinity possible spaces (discrete or continuous) each of them could be created, based on, some fundamental properties. However, we will restrict our analysis to spaces that are closely related to the Euclidean space which could be introduced by utilizing the concept of curvature. One manifestation of this concept is the intrinsic geometrical difference between curves and straight lines. These geometrical differences, captured by that concept, can be generalized to spaces with higher dimensions (two, three, four, etc.). However, for simplicity, a constant curvature is assumed (i.e., homogenous and isotropic spaces). It can be shown that in this case the curvature of a space is given by "Theorema Egregium" which was deduced by the mathematician Carl Gauss. We will state below this theorem without proof for the special case of two dimensional space:

$$k = \frac{1}{2g_{11}g_{22}}\left\{\frac{1}{2g_{11}}\left[\frac{\partial g_{11}}{\partial x^1}\frac{\partial g_{22}}{\partial x^1} + \left(\frac{\partial g_{11}}{\partial x^2}\right)^2\right]\right.$$

$$+\frac{1}{2g_{22}}\left[\frac{\partial g_{11}}{\partial x^2}\frac{\partial g_{22}}{\partial x^2}+\left(\frac{\partial g_{22}}{\partial x^1}\right)^2\right]$$

$$\left.-\left[\frac{\partial^2 g_{11}}{\partial (x^2)^2}+\frac{\partial^2 g_{22}}{\partial (x^1)^2}\right]\right\} \quad (1.13)$$

As an application of this theorem to two-dimensional space, we notice that since the sphere is a two-dimensional surface embedded in the Euclidean space, its metric is given by the previously deduced metric as a special case, namely:

$$ds^2 = R^2 d\theta^2 + R^2 \sin^2\theta d\phi^2 \quad (1.14)$$

and hence the metric tensor that describes the intrinsic geometry of its surface is:

$$g_{ij} = \begin{bmatrix} R^2 & 0 \\ 0 & R^2\sin^2\theta \end{bmatrix} \quad (1.15)$$

so we have:

$$g_{11} = R^2, \quad g_{22} = R^2\sin^2\theta, \quad x^1 = \theta, \quad x^2 = \phi$$

from that it follows:

$$\frac{\partial g_{11}}{\partial x^1} = \frac{\partial g_{11}}{\partial x^2} = \frac{\partial g_{22}}{\partial x^2} = 0$$

so

$$k = \frac{1}{2g_{11}g_{22}}\left[-\frac{\partial^2 g_{22}}{\partial (x^1)^2}+\frac{1}{2g_{22}}\left(\frac{\partial g_{22}}{\partial x^1}\right)^2\right]$$

and hence:

$$\begin{aligned} k &= \frac{1}{2R^4 \sin^2\theta}\left[2R^2(\sin^2\theta - \cos^2\theta) + \frac{4R^4 \sin^2\theta\cos^2\theta}{2R^2\sin^2\theta}\right] \\ &= \frac{1}{R^2}\left[1 - \frac{\cos^2\theta}{\sin^2\theta} + \frac{\cos^2\theta}{\sin^2\theta}\right] \\ &= \frac{1}{R^2} \end{aligned} \tag{1.16}$$

This shows that the curvature of a spherical surface is the reciprocal of the square of its radius, so as the size of the sphere and its radius approach infinity, its geometry approaches the geometry of the flat plane that has a zero curvature which is in a perfect consistence with observation. This space is an example of a two-dimensional Riemann space with a positive constant curvature. To describe this space in a more convenient way, we will introduce the following coordinate system (see Figure 1.1 below):

$$x = ar\cos\phi, \quad y = ar\sin\phi, \quad z = a\sqrt{1-r^2} \tag{1.17}$$

where r is a dimensionless parameter defined by $r = \sin\theta$ and so its values are contained in the interval [0,1]. Therefore by noticing that:

$$r = \sin\theta \Rightarrow \sin^{-1} r = \theta \Rightarrow d\theta = \frac{1}{\sqrt{1-r^2}} \tag{1.18}$$

it follows from the above derived metric of the sphere that it has in this new coordinate system the following form:

$$ds^2 = a^2\frac{dr^2}{1-r^2} + a^2r^2d\phi^2 \Rightarrow g_{ij} = \begin{bmatrix} \frac{a^2}{1-r^2} & 0 \\ 0 & a^2r^2 \end{bmatrix} \tag{1.19}$$

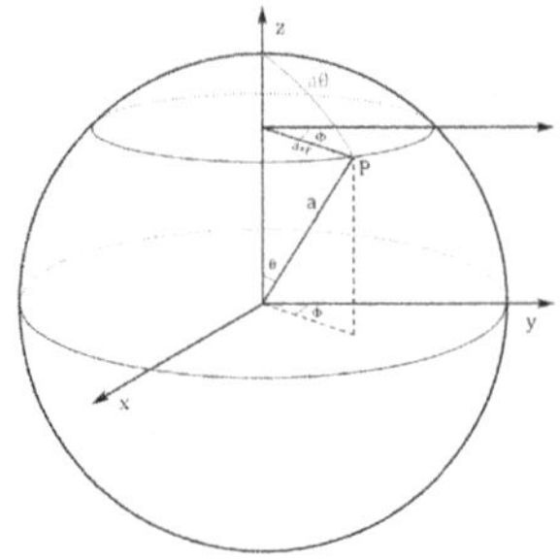

Figure 1.1: The spherical plane embedded in the Euclidean space along with the adopted modified spherical coordinates.

Another example of non-Euclidean two-dimensional surface is the hyperbolic plane (Figure 1.2). This plane is embedded in the pseudo-Euclidean space which has the metric tensor:

$$g_{ij} = \begin{bmatrix} 1 & 0 & 0 \\ 0 & 1 & 0 \\ 0 & 0 & -1 \end{bmatrix} \tag{1.20}$$

and hence the metric:

$$ds^2 = dx^2 + dy^2 - dz^2 \tag{1.21}$$

so, using equation (2.8), it is a straightforward to derive the metric tensor of this space in another disguise based on the following coordinate system:

$$x = a \sinh\theta \cos\phi, \quad y = a \sinh\theta \sin\phi, \quad z = a \cosh\theta \tag{1.22}$$

if we let $r = \sinh\theta$ to produce a new coordinate system and use the definition of the hyperbolic plane as the set of points in a pseudo-Euclidean space which has a fixed distance (according to its metric) from a specific point, we would derive the following metric and fundamental tensor of that surface:

$$ds^2 = a^2 \frac{dr^2}{1+r^2} + a^2 r^2 d\phi^2 \Rightarrow g_{ij} = \begin{bmatrix} \frac{a^2}{1+r^2} & 0 \\ 0 & a^2 r^2 \end{bmatrix} \quad (1.23)$$

now since the Euclidean plane has in polar coordinates the metric tensor:

$$g_{ij} = \begin{bmatrix} 1 & 0 \\ 0 & r^2 \end{bmatrix} \quad (1.24)$$

we could combine the preceding results and have the spherical, hyperbolic and Euclidean planes as special cases of a general tensor by letting K=(-1,0,+1) in the following metric tensor:

$$g_{ij} = \begin{bmatrix} \frac{a^2}{1-Kr^2} & 0 \\ 0 & a^2 r^2 \end{bmatrix} \quad (1.25)$$

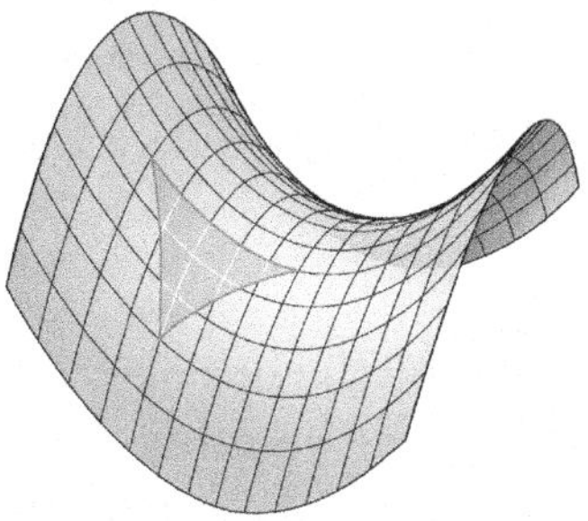

Figure 1.2: A simplified depiction of the hyperbolic plane.

the constant K which is defined in the preceding way is actually the curvature of the surface in each case. This result can be proven using equation (2.12) as follows:

$$g_{11} = \frac{a^2}{1-Kr^2}, \quad g_{22} = a^2 r^2, \quad x^1 = r, \quad x^2 = \phi$$

which entails that

$$\frac{\partial g_{11}}{\partial x^2} = \frac{\partial g_{22}}{\partial x^2} = 0$$

so we have:

$$k = \frac{1}{2g_{11}g_{22}} \left[\frac{1}{2g_{11}} \left(\frac{\partial g_{11}}{\partial x^1} \frac{\partial g_{22}}{\partial x^1} \right) + \frac{1}{2g_{22}} \left(\frac{\partial g_{22}}{\partial x^1} \right)^2 - \frac{\partial^2 g_{22}}{\partial (x^1)^2} \right]$$

and hence:

$$\begin{aligned} k &= \frac{1 - Kr^2}{2a^4r^2} \left[\frac{1 - Kr^2}{2a^2} \times \frac{2a^2rK}{(1 - Kr^2)^2} \right. \\ &\qquad \left. \times\, 2a^2r + \frac{1}{2a^2r^2} \times 4a^4r^2 - 2a^2 \right] \qquad (1.26) \\ &= \frac{1 - Kr^2}{2a^4r^2} \times \left[\frac{2a^2r^2K}{1 - Kr^2} + 2a^2 - 2a^2 \right] \\ &= \frac{K}{a^2} \end{aligned}$$

which is the curvature of the surface in each case. Following a similar reasoning, we could generalize these concepts to three-dimensional space leading to the following metric tensor:

$$g_{ij} = \begin{bmatrix} \frac{a^2}{1-Kr^2} & 0 & 0 \\ 0 & a^2r^2 & 0 \\ 0 & 0 & a^2r^2 \sin^2\theta \end{bmatrix} \qquad (1.27)$$

However, the relativistic phenomena that arose as logical consequences of experimentally confirmed assumptions behave as if our three-dimensional space, which is the scene of the observable events, embedded and moved with speed of light in a four-dimensional pseudo-Euclidean space called "Minkowski space". This space has, in Cartesian coordinates, the metric tensor:

$$g_{ij} = \begin{bmatrix} 1 & 0 & 0 & 0 \\ 0 & -1 & 0 & 0 \\ 0 & 0 & -1 & 0 \\ 0 & 0 & 0 & -1 \end{bmatrix} \tag{1.28}$$

the events in this space are 4-tuples $x^{\mu} = (ct, x, y, z)$ and hence its metric, given in that coordinate system, has the form:

$$ds^s = c^2dt^2 - dx^2 - dy^2 - dz^2 \tag{1.29}$$

if we adopt this experimentally supported view, the metric tensor of our pseudo-Riemannian space (the space-time continuum) would have, in our amended spherical coordinates, the form:

$$g_{ij} = \begin{bmatrix} 1 & 0 & 0 & 0 \\ 0 & -\frac{a^2}{1-Kr^2} & 0 & 0 \\ 0 & 0 & -a^2r^2 & 0 \\ 0 & 0 & 0 & -a^2r^2\sin^2\theta \end{bmatrix} \tag{1.30}$$

and hence the following metric:

$$ds^2 = c^2dt^2 - a^2(t)\left[\frac{dr^2}{1-Kr^2} + r^2(d\theta^2 + \sin^2\theta d\phi^2)\right] \tag{1.31}$$

This metric is called "Robertson-Walker metric" and it predicts and explains many cosmological phenomena which have been proven to be consistent with observations. Three types of hypothetical universes can be conceived through this metric depending on the density of matter in the universe. Those types are the closed, flat and open universes, and each of these universes has its own geometry. The closed universe has positive curvature and hence spherical geometry so that in this type, the parallel lines converge to a point, the sum of the angles in a geodesic triangle is more than 180°, the circumference of a circle divided by its diameter is less than π and the surface area of a sphere as a function of its radius is periodic. The flat universe, which has zero curvature, is supplied with Euclidean geometry, so the parallel lines have a fixed vertical distance between them, the geodesic triangles have a total sum of their interior angles of 180°, the ratio of the circumference of a circle to its diameter is equal to π and the surface area of a sphere is proportional to its radius. Finally, the open universe with a negative curvature has a hyperbolic geometry in which the parallel lines diverge, the sum of the interior angles of a geodesic triangle is less than 180°, the circle circumference and its diameter have a ratio greater that π and the surface area of the sphere and its radius vary exponentially. The previous geometrical differences between these types of universes are just examples of some aspects of the differences between their geometries. As a result of that, the relations and aspects of the geometrical elements in each universe differ widely, leading to the requirement of different capability to appreciate them. Therefore, the inherent faculty of the creatures, in this universe, to conceive the geometrical objects, their symmetrical aspects and relations and hence to appreciate the relative distances, sizes and motions was a reflection of their development in the Euclidean space which would be different if they were developed in other geometrical environments. This fact emphasizes the relativistic nature of space and its dependence on the constitutional structure of the observer. However, to appreciate the extent of this dependence, we notice that those three types of universes can be treated as special cases of a more general space that has infinitely many other special cases of discrete and continuous spaces with any variety of geometrical aspects that can be conceived. This assumption opens the door to infinitely many possibilities in which a space could be perceived depending on the parent geometrical environment of the observer.

Order and chaos are two faces of the same coin. In fact, one of the main functions of science is to organize chaos. If we have two sets of points A and B (see Figure 1.3) and we want to derive the set of rules which describe, quantitatively, their distances from an arbitrary line, the points of the set A will be described by a single rule, but each point in the set B will be described by a special rule. We consider the system A as a well organized system and the system B as a very chaotic system, so the increase of the degree of chaos in a given system is a reflection of the increase of the order of the set of rules required to describe that system. However, the fuzzy boundary line between the systems which are designated as chaotic systems and those described as ordered systems depends on the constitutional structure of the observer. (That boundary line is not the same for chimpanzees and human being.)

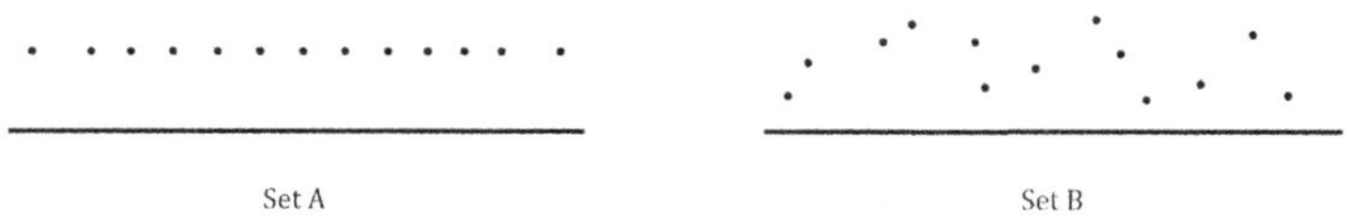

Figure 1.3: An organized system (set A) and a chaotic system (set B).

Our intellectual judgment of whether a particular behavior is right or wrong depends on our principles, customs, values and traditions, so these constituents compose the intellectual frame of reference that determines the natures of the correct and wrong actions. As a consequence of this fact and if we allow each of these constituents to have any possible form, the virtues and vices would be relativistic phenomena that depend on the experience of the observer, and hence there are no absolute rightness and wrongness. However, the existence of common universal virtues and vices in our world is a reflection of the similarities in the constituents of our intellectual frame of references, so we agree on the correctness of the behaviors that are based on these similarities and disagree elsewhere. This observation entails the dependence of the nature of any behavior on the experimental constitutional structure of the observer that is shaped by the intellectual environment in which he has developed.

The previous arguments inductively support the relativistic natures of what we receive by our senses and judge in our thoughts. The natures of the physical world around us and the intellectual thoughts in our brains were created by our constitutional structures which are mirrors of the physical and intellectual environments in which we have evolved. The metric of the four-dimensional pseudo-Reimannian space, which its spatial part is the scene of the observable events, emphasizes the relativistic natures of past, present, future and the essence and appreciation of time. The dimensions of the ultimate constituents (the elementary particles) of our bodies and our environmentally shaped senses determined the identity of the physical world around us. Our similarities in the intellectual experience and psychological structure produce a common sense of virtues and vices. These support the proposition that the identity of any concept depends on the constitutional structure of the observer and if we relate that concept to a suitable static or dynamic constitutional structure, it "theoretically" would have any prescribed entity. This line of thought emphasizes that the constituents of these entities have relativistic properties that depend on the structure of the observer. This phenomenon is not restricted to the preceding examples. Actually, careful reasoning based on the accumulated human endeavors to explore the living reality inductively supports the existence of this phenomenon as a ubiquitous aspect of the characters of any entity.

The only intrinsic nature of any meaning is that it is a meaning, so we will use the word "concept" to refer to any abstract or concrete meaning. The constituents of any concept are again concepts which in turn consist of concepts, so any concept is a fractal by its nature. On the other hand, any set of concepts has intimately real or imaginary extension. We call this set of concepts a "reality" and its extension a "science". For instance, the previous sentence is a set of graphs (concepts) and it has a set of geometrical spacing, symmetries and shapes as an extension, so this set of graphs constitutes a reality and its geometrical aspects (its extension) constitutes a science. In fact, this relation is reversible, that is, each of the previous set of concepts is an extension of the other and hence a reality and science of each other. We call any set of concepts and its extension (any reality and its science) an "existence". However, since any group of concepts, which may be considered a reality or science, form collectively a new concept, any concept is a reality and science. Moreover, since any set of concepts, which form collectively a concept, can be divided into two subsets each of them is an extension of the other, from real or imaginary respects, any concept is also an existence. However, the concept can be treated as a property of another concept, from some respects, hence any concept is, also, an attribute. For example, if we have a red box, the red color is a concept, reality, science and existence, but it also is an attribute of the box. This treatment can be generalized to any abstract or concrete meaning that can be conceived.

1.2 Concepts

Aristotle has talked about the upper ten classes of meanings (known as the ten categories) in what is called today the material logic. This classification was apparently based on the assumption that any class of meaning has an absolute essence which is independent from the constitutional structure of the observer. However, as we have shown, the accumulated knowledge in many disciplines strongly supports that the essence of any meaning is a relativistic phenomenon which depends on the observer. This entails that the ten categories do, in fact, describe the essences of meanings with respect to the constitutional structure of humans. In this monograph, we intend to develop a new conceptual system to treat these concepts more abstractly, so we classify meanings according to their receptive entities rather than their absolute essences which, actually, do not exist. If we define "intelligence" to be the ability of processing data and producing knowledge, we could distinguish between the insulated and non-insulated intelligence. The first type has no connections with its exterior world and hence can not be considered as an observer, while the second type is connected to the outside world, so any observer is, by nature, a non-insulated intelligence. To achieve communication with the exterior world, the observer must have a probe to detect the mediator of the data, we call this probe a "sense". Theoretically, the observer can have as many senses as his structure allows him to differentiate different meanings. As a result of the conditions in our terrestrial developing environment, we have earned five senses that enable us to classify sensible meanings into five types (image, sound, feel, smell and taste). However, the sensation of these different meanings and the interaction with the surrounding environments, in the existence of the intellectual activity, create two internal domains that produce conscious and emotion. As a consequence of that, two more types, essentially different from the five sensational concepts, originate. We call these two types "thought" and "affection", respectively. These meanings constitute the "orthodox meanings" and the general observer, that has developed those five senses and their ramifications to distinguish between the latter seven types of meaning, is called the "orthodox observer".

Each of the orthodox meanings has genuine, intermediate and mental aspects. We may use the concept of representation to distinguish between them. The genuine aspect of a meaning depicts the representation of the concept by itself while the intermediate aspect describes its representation by another concept and the mental aspect is the observer's intellectual representation of the concept. If we take the tree as an example, the tree itself is a genuine aspect of itself, the word of tree is an intermediate aspect of it and the intellectual idea of the tree, with respect to an observer, is its mental aspect. There is just one genuine aspect for any orthodox meaning, but infinitely many intermediate and mental aspects for it. However, the type of any intermediate aspect of a concept is one of the seven types of the orthodox meanings, so there are visual, audible, tactile, gustatory, smelled, affective and mental intermediate aspects. For instance, the written word of the sun is a visual intermediate aspect of the actual sun while its pronunciation is an audible intermediate aspect of it. The first two types (visual and audible) of the intermediate aspects are widely used, while the third type (tactile) is rarely utilized (braille language), but the rest of these intermediate types are obsolete. The word "name" is the common word for the intermediate or the middle aspect. Compactness and pureness are properties of the middle aspects (names), so some middle aspects are more elaborated than others and some of them purer than others. The equation of a straight line, for example, is more compact than its graph, and both of them are indirect visual middle aspects (names) of the relation which is genuinely a mental concept. However, the figure "2" is a direct (pure) visual intermediate aspect of a mental concept of a quantity. This discussion and distinctions are observed in any instance where we encounter intermediate aspects or names. It should be emphasized that there are no absolute essences of these three aspects, but their essences are relativistic phenomena that depend on the constitutional structure of the orthodox observer.

2 Sciences

As we have mentioned, any science is an extension of a reality and one of the five natures of any concept is a science. The commutative natures of science and reality, such that one of them is an extension of the other, lead to the fact that science and reality are two faces of the same coin. For instance, medicine, in its genuine aspect, is a science of human body and human body is a science of medicine, and each of them is a reality of the other. Therefore, the assertion that a particular set of concepts is a science of a real or imaginary extension, which constitutes its reality, is a matter of definition that reflects our scientific objects. This means that the environments of the scientific objects determine the sets of concepts which form the extensions of the realities and hence are considered as sciences. The natures of human interests produce objective environments that coincide with the sensibles, so the latter are taken as realities and their extensions as "human sciences". As a result, human sciences can be defined as the extensions of the sensibles. If we return to the previous example of medicine and human body, we found that the concepts which constitute the human body are sensibles, so they are taken as a reality and their extension (medicine) as a human science. The relative "stability" of the concepts that collectively compose a reality has a transitive effect on the stability of the concepts that constitute its science, so the "real" science mimics the instability of its reality. However, if science is developed without allowing for this property of its reality, the phenomenon of accurate and inaccurate sciences would be produced as a reflection of different degrees of instability. For example, psychology (the science of affective structure) is less accurate than physics (the science of nature) because the properties of the four physical constituents of the universe (matter, energy, space and time) and their relations are much more stable than the constituents of the human psychological structure, which varies widely between humans. (This also explains why medicine is more accurate than psychology.) Another property of the reality which is related to the stability and has a reflection on its extension or science is "uniformity". This property of a reality determines the degree of "complexity" of its science, so the more uniform a reality, the more simple its science. Theoretically, any reality with any degrees of stability and uniformity has a science. However, for the limited capacity at a given time, the realities with high degrees of instability and non-uniformity are considered as random systems and treated by inferential statistics and other methods involving uncertainty principles such as quantum mechanics and quantum field theory. For a given reality, the "rational data", obtained by observations and experiments, determine its extension or science. The rational data can be defined as the collection of data that realizes or facilitates the realization of the scientific objects. Therefore, we may accurately define the science as a set of correlated and isolated concepts that are connected to the rational data and arise from the relative stability of the environment of the scientific objects. It follows from this definition of science that "any" object has a science, so if someone would like to drive a car, there is a science to accomplish this task. The reality of this science is the structure of the car and the conditions in its environment. On the other hand, if a nation would like to build a colony on the moon, there are sciences for these realities to achieve this object (physics, space sciences, engineering and architecture). Even for the aim to win a backgammon game, in which its structure and rules constitute a reality, there is an extension of this reality that constitutes its science to accomplish this objective. The definition of science as a particular set of concepts and the existence of genuine, intermediate and mental aspects of any concept lead to the corresponding three aspects of the science. We call its middle aspect (name) the "Adamic science".

2.1 Adamic sciences

Based on its previous definition as an intermediate aspect, there are an infinitely many possible Adamic sciences or names for the extension of given sensibles as a reality. However, for simplicity, sufficiency and historical reasons, only a few of them were utilized. Henceforth, we will drop the word "Adamic" and refer to those intermediate aspects just as sciences. We may distinguish between "perfect" and "imperfect" sciences according to the degree of the assignment of their realities. If the reality of a science was completely assigned, then the science is perfect. Otherwise it is considered imperfect, so chemistry or any natural science is imperfect, while the science of chess or any game is a perfect science. Also, for educational reasons, the realities of sciences are simplified with different degrees of simplifications which produces sciences of different complexity for a given reality. This kind of simplification is pronounced when comparing the sciences taught in the fundamental education with the same sciences taught at the university level and the latter with them in the postgraduate level. It is obvious that the simplicity of the same reality decreases with the educational level and, as a consequence, the science complexity increases. Another property of science, that arises from the limited capacity of the observer, is the scientific branching. This phenomenon occurs when a huge accumulated data, that outlines the science, revolves about two or more distinguishable scientific areas, so this problem is mitigated by splitting them, based on these areas, into several sciences. For instance, statistics was the branch of mathematics that deals with the description of data and the scientific principles of taking decisions evolving in-deterministic experiments. However, for the stated reason, they have been split into two distinct sciences.

2.2 The constituents of science

Science, as any concept, arises from the relative stability of a set of constitutional concepts, so its nature and behavior depend on these concepts. Language and definitions are constituents of any science. However, quantitative sciences have the relations, measurand systems and unit systems as three additional constitutional concepts. We thoroughly analyze these constituents below.

2.2.1 Language

The human brain, as an example of the intellectual processor of the orthodox observer, is made of nervous cells that create intellectual domains by sending nervous signals to each other. The genuine aspects of these domains are physical by nature and very complicated, so the simplifying and organizing method that was acquired through the natural development of the human race was the language. However, since the nervous signals are the responses of the mental cells to the stimulation from the surrounding environment, the intellectual domains have developed in such a way to reflect a relative essential nature of this environment, so the physical aspects of the intellectual domains mirror the chronological feature and the distinction between different meaning that exist in their surroundings. As a ramification, the language, as a simplifying and organizing medium to the intellectual domains, should have these features. In language, the temporal feature is captured by the concept of "tenses", and the sense of three tenses (past, present and future) naturally appeared in the human language. However, there are some meanings that have chronological aspects while others do not essentially have the temporal tincture and still others do not have "stand-alone" meanings, so the concepts of verb, noun and preposition have developed. The language is the collection of these concepts and their rational constructions (syntax) that produce logic and sensible reasoning. Technically, there are infinitely many possibilities of the middle aspects (names) of the linguistic concepts and their syntaxes. (This fact manifests itself in the thousands of languages that exist in the world.) However, in general, the only used of the seven types of the middle aspect in any human language are the visual and the audible aspects (except some cases in which tactile aspect is used for blinds). Therefore language provides science with the visual and audible aspects of the linguistic concepts and their rational structure (syntax) that contains the logical and sensible reasoning. Any tool that has been developed by an orthodox observer to perform the preceding functions is called an "orthodox language" or in short a language if there is no confusion.

2.2.2 Definitions

Concept is one of the five faces of any meaning, and the constitutional structure of the meaning as a concept is created by definition, so the existence and the nature of a concept depend on the existence and the features of its definition. This ontological relation between concepts and their definitions necessitates the existence of the structural concepts in the suitable correlations and configurations that set up the definitions and hence collectively create the concepts. The prior existence of the definition to its concept is a privileged property of the genuine aspect of the definition, so its middle (name) and mental aspects succeed the existence of the concept. For example, the genuine aspect of the definition of the horse or its constructing concepts (flesh, bones, organs, etc.) and their relations occur before the existence of the genuine aspect of the concept of horse, since they are the constitutional structure of that concept. However, the intermediate aspect (in the dictionary) and the mental aspect (in human mind) succeed its existence. The mental concept as one of the seven types of the orthodox meanings has, by its nature, the distinctive character that the creation of any possible middle aspect of its definition entails the existence of its genuine aspect. As one manifestation of this phenomenon, the linguistic definition (name) of the cross product of vectors produces the genuine mental aspect of its definition. However, the rest of the orthodox meanings do not have this privilege. For instance, the creation of an intermediate aspect of the definition of the Phoenix in a particular language does not lead to the existence of its genuine aspect in reality. In the latter case the concept is called "fictitious". This property of the mental aspect arises from the capability of the orthodox observer to construct the intellectual concept according to a prescribed intermediate aspect of its definition, so the other orthodox meanings would have the same privilege if the orthodox observer was endowed with the capability to construct the genuine aspects of the created "names" of their definitions. This effect is demonstrated by the construction of a car based on its artistic depiction and engineering design which constitute the intermediate aspect of its definition. In principle, there are countless possibilities of definitions that can be used to perform a particular function which leads to the creation of a distinctive intellectual structure. However, for simplicity and uniformity, only limited flavors of them are employed.

However, the constitutional structure created by the genuine aspect of a definition may exist in more than one entity. This produces a class of entities with the characteristics created by that definition. We call this class the "extension" of the definition. The uncertainty in determining the extension of a given definition produces a confusion about the exact entities that are included in the extension. This confusion is called the "vagueness" of the definition. Another important phenomenon, inherent in any language, is the use of the same middle aspect (name) to represent different meanings. This effect creates the so called the "ambiguity" of the definition. To overcome these problems, we always use "stipulated definitions" for the linguistic intermediate aspects used in science which should be partially descriptive (have related meanings to their meanings in the language). The latter requirement is essential since it reduces the insanity results from the confliction between the stipulated and descriptive meanings. We call any possible collection of partially stipulated definitions that can be used in a particular science a "nomenclature".

There are, in general, two main types of definitions the "intentional definition" and the "extensional definition". The former type defines the concept by its constituents and hence the existence of the genuine aspect of that definition creates the constitutional structure of that concept. However, the extensional definition defines a concept by determining its extension, so the existence of its genuine aspect entails the existence of the genuine aspect of its extension. The vagueness associated with the uncertainty in determining the extension in the second type of definition can be eliminated by naming that extension if its members are finite and few. Otherwise the extension can be determined recursively or operationally. For instance, we may define, recursively, a set of "parallel lines" in the Euclidean plane as the set parallel to a particular line, so we could use this line or any other line (parallel to it) to construct that set. On the other hand, we may define, operationally, the "excellent student" as the student who achieves more than a particular score of a given test.

2.2.3 Relations

Similarity is a concept that describes a common attribute among several orthodox meanings, so this concept intellectually isolates a number of meanings. We call this intellectual isolation of assemblage of meanings a "set" and its members "elements" or "points". It should be emphasized that the common attribute could be "any" descriptive meaning including the membership to that assemblage by definition. As a consequence, the set could be assigned by stating explicitly its members or by stating the common attribute that creates the intellectual isolation or the set. For example, the set $S_1 = \{1, 7, 5, 9\}$ is assigned by stating its elements or points, and the set $S_2 = \{$x such that x is the capital of a county in the world$\}$ is assigned by its attribute. For brevity, we use the symbols $\in$ and $\notin$ to indicate, respectively, that a given element is a member or not a member of a particular set, so for the previous sets S_1 and S_2, we have $5 \in S_1$ and Rome $\notin S_1$ and the situation is exactly the opposite for S_2. However, any element does, in fact, constitute a set by itself, so we may generalize the concept of membership by introducing the symbols $\subseteq$ and $\nsubseteq$ to indicate, respectively, that a particular set is included and not included in another set. Moreover, if the set is perfectly included (the inclusion is not commutative), we call the included set a "proper set" of the comprising one and we use the symbol $\subset$ to indicate this case. The set may have a limited number of elements (such as S_1 or S_2) in which case it is called a "finite set". Otherwise it is called an "infinite set". (The set of integer numbers can be taken as an example of the infinite set.) If a set S is a finite set, we will refer to the number of its elements by $|S|$.

The elements of a set, as we have mentioned, have at least one common attribute that entails the intellectual isolation which creates their existence as a set. However, some other common attributes may also exist for the entire or part of the members of the set. This phenomenon can be described by introducing the concept of "quantifier", but since we are only interested in determining whether the entire or part of a set is intellectually isolated by a given attribute, we only use the "universal" and "existential" quantifiers and refer to them with the respective symbols $\forall$ and $\exists$. For instance, the attribute that a number is a prime intellectually isolates the numbers 5 and 7 from set S_1, so this can be described symbolically as follows "$\exists\ x \in S_1 \ni P_1(x)$" which reads "there is elements in S_1 such that (referred to by the symbol $\ni$) the proposition $P_1(x)$, which states that x is a prime number, is true". On the other hand, the attribute that a number is odd intellectually isolates the whole members of the set S_1. This has the symbolic description "$\forall\ x \in S_1 : P_2(x)$" which reads " for all x an element of S_1, the proposition $P_2(x)$ (which states the oddity) is true". We call the set which is composed of all the possible subsets of a given set S the "power set" of the set S and denote it by $p(S)$. As an example, the power set of the set $S_3 = \{2, 3, 4\}$ is the following set:

$$p(S_3) = \{\phi, \{2\}, \{3\}, \{4\}, \{2,3\}, \{2,4\}, \{3,4\}, S_3\} \tag{2.1}$$

where ϕ refers to the empty set. Based on the relevant definitions, it can be shown that the number of the elements of the power set of a set S which has n elements is $|p(S)| = 2^n$. Applying this to the power set $p(S_3)$, we find $|p(S_3)| = 2^3 = 8$ which is, actually, the number of elements (subsets) that are included in $p(S_3)$. Utilizing appropriate notations, the basic two algebraic operations of sets (union and intersection) can be defined, respectively, for any two given sets A and B as follows:

$$A \cup B = \{x|(x \in A) \vee (x \in B)\} \tag{2.2}$$
$$A \cap B = \{x|(x \in A) \wedge (x \in B)\} \tag{2.3}$$

with the respective generalization:

$$\begin{aligned} \bigcup_{i=1}^{n} A_i &= A_1 \cup A_2 \cup ... \cup A_n \\ &= \{x|(x \in A_1) \vee (x \in A_2) \vee ... \vee (x \in A_n)\} \\ &= \{x|\ \exists\ i \ni x \in A_i, n \geq i \geq 1\} \end{aligned}$$

and

$$\begin{aligned} \bigcap_{i=1}^{n} A_i &= A_1 \cap A_2 \cap ... \cap A_n \\ &= \{x|(x \in A_1) \wedge (x \in A_2) \wedge ... \wedge (x \in A_n)\} \\ &= \{x|\ \forall\ i : x \in A_i, n \geq i \geq 1\} \end{aligned}$$

two other operations (difference and symmetric difference) can be defined as follows:

$$A - B = \{x|(x \in A) \wedge (x \notin B)\} \tag{2.4}$$
$$A \Delta B = \{x|(x \in A \wedge x \notin B) \vee (x \in B \wedge x \notin A)\} \tag{2.5}$$

where $\cup, \cap, -, \Delta, \vee$ and $\wedge$ have the respective indications "union", "intersection", "difference", "symmetric difference", "or" and "and". The universal set Ω of two sets A and B is defined as the union of these two sets. This notion creates a mutual relation between those sets with respect to Ω. We call this relation complementary and define it as follows:

$$\mathrm{B} = \mathrm{A'} = \{x|(x \in \Omega) \wedge (x \notin \mathrm{A})\} \tag{2.6}$$

here the "prime" refers to the complement of the set A. As a generalization of this concept, any set Ω that has the following property:

$$\bigcup_{i=1}^{n} A_i \subseteq \Omega \tag{2.7}$$

with respect to a collection of sets $A_1, A_2, ..., A_i$ is called the universal set of those sets. By definition, changing the order of the elements in a given set does not produce a new set, so we may state its elements in any order and we still have the same set. However, for its important application, we may capture the concept of ordered assemblage of elements by using special types of sets, so the ordered sequence of elements (n-tuple) can be defined as follows:

$$\begin{aligned}(a_1, a_2, ..., a_n) &= \{a_1, \{a_1, a_2\}, \{a_1, a_2, a_3\}, \\ &\quad ..., \{a_1, a_2, ..., a_n\}\}\end{aligned} \tag{2.8}$$

This definition preserves the elements order of the n-tuple. We will illustrate that for the special case $n = 2$. Suppose that (a_1, a_2) and (b_1, b_2) are two ordered pairs (2-tuple) then by definition we have:

$$\begin{aligned}(a_1, a_2) = (b_1, b_2) &\Rightarrow \{a_1, \{a_1, a_2\}\} = \{b_1, \{b_1, b_2\}\} \\ &\Rightarrow a_1 = b_1 \wedge \{a_1, a_2\} = \{b_1, b_2\} \\ &\Rightarrow a_1 = b_1 \wedge a_2 = b_2\end{aligned}$$

so as a result if $(a_1, a_2, ..., a_n)$ and $(b_1, b_2, ..., b_n)$ were two n-tuples then:

$$(a_1, a_2, ..., a_n) = (b_1, b_2, ..., b_n) \Leftrightarrow a_i = b_i : \forall\, i \tag{2.9}$$

We define the "Cartesian product" of a number of sets A_1, $A_2, ..., A_n$ as the n-tuples that have the following construction:

$$\begin{aligned} A = \prod_{i=1}^{n} &= A_1 \times A_2 \times ... \times A_n \\ &= \{(x_1, ..., x_n) | (x_1 \in A_1) \wedge ... \wedge (x_n \in A_n)\} \\ &= \{(x_1, ..., x_n) | x_i \in A_i : \forall\, i \in I = \{1, 2, ..., n\}\} \end{aligned}$$

and, for brevity, in the case that $A_1 = A_2 = ... = A_n$, the notation $A^n = A_1 \times A_2 \times ... \times A_n$ will be utilized. For example, if we have, as a special case, the two sets $A_1 = \{a, b, c\}$ and $A_2 = \{d, e, f\}$ then their Cartesian product will be:

$$\begin{aligned} A_1 \times A_2 = \{&(a, d), (a, e), (a, f), (b, d), (b, e), \\ &(b, f), (c, d), (c, e), (c, f)\} \end{aligned} \tag{2.10}$$

notice that, by definition, $A_1 \times A_2 \neq A_2 \times A_1$. The Cartesian products of sets can be represented using graphs, tables or arrow diagrams. These three types of representation are illustrated in Figure 2.1 below for the Cartesian product $A_1 \times A_2$ of the previous two sets.

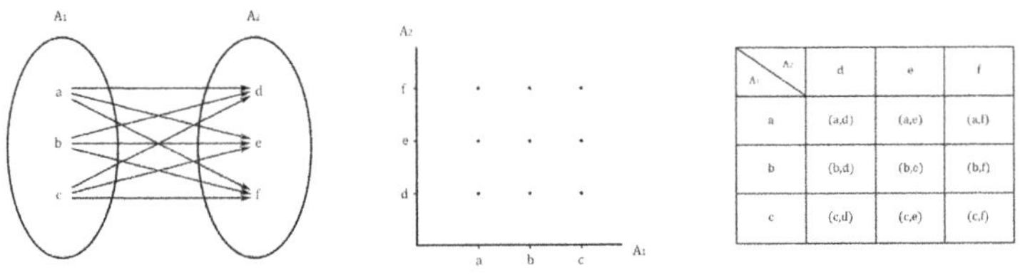

Figure 2.1: The graphical, tabular and arrow representation of the Cartesian product $A_1 \times A_2$.

It is straight forward to prove that if a set B_1 is a subset of another set A_1 and a set B_2 is a subset of a set A_2 then the Cartesian product $B_1 \times B_2$ is included in $A_1 \times A_2$. We call the subset $R = B_1 \times B_2$ a "binary relation" from A_1 to A_2, so if $(x, y) \in R$, we say x related to y by R and refer to that by the notation $x \; R \; y$. Otherwise we write $x \not R \; y$. In the case that $A_1 = A_2 = A$, we say R is a binary relation on A^2. If R is a binary relation from a set A to a set B, we call the set A and the set B the "departure" and the "destination" of R, respectively. On the other hand, the respective subsets from A and B defined by $\{x|x \in A \wedge xRy\}$ and $\{y|y \in B \wedge xRy\}$ are referred to as the "domain" and the "range" of R and their union is called collectively the field of R. Using the appropriate Cartesian products, we may generalize this concept of relation to any given number of elements. However, for brevity, we will use the word "relation" to refer to the binary relation unless otherwise stated. As an illustration of this concept, the Euclidean plane and space are the respective graphical representations (visual intermediate aspects) of the Cartesian products $\mathbb{R}^2 = \mathbb{R} \times \mathbb{R}$ and $\mathbb{R}^3 = \mathbb{R} \times \mathbb{R} \times \mathbb{R}$ of the set of real numbers and hence, based on the previous proposition, any curve or shape in these spaces constitutes a representation (middle aspect) of a subset of the proceeding Cartesian products of $\mathbb{R}$ and therefore an intermediate aspect of a relation on $\mathbb{R}^2$ and $\mathbb{R}^3$ respectively. We denote the inverse relation of a given relation R by R^{-1} and define it as:

$$R^{-1} = \{(y, x)|(x, y) \in R\} \tag{2.11}$$

The reflexivity, symmetry and transitivity are important properties of the relations that are defined on a given set A. A relation R (defined on A) is described as "reflexive", "symmetric" or "transitive" if the respective attributes: $xRx : \forall x \in A$, $(x, y) \in R \Rightarrow (y, x) \in R$ or $(x, y), (y, z) \in R \Rightarrow (x, z) \in R$ are realized for that relation. If all of these attributes are applicable to the given relation, we call it an "equivalence relation" on A. For instance, if we consider the parallelism relation on the Euclidean space $\mathbb{R}^3$, we find that any three lines l, l' and l'' in $\mathbb{R}^3$ satisfy the following conditions:

$$l \parallel l : \forall \, l \in \mathbb{R}^3, \;\; l \parallel l' \Rightarrow l' \parallel l \tag{2.12}$$

and

$$l \parallel l' \wedge l' \parallel l'' \Rightarrow l \parallel l'' \tag{2.13}$$

hence, this relation constitutes an equivalence relation on $\mathbb{R}^3$. However, the perpendicularity relation on the Euclidean plane $\mathbb{R}^2$ is not an equivalence relation, since any line in $\mathbb{R}^2$ is not perpendicular to itself (the relation is not reflexive) and if a line is perpendicular to two lines in $\mathbb{R}^2$, this does not entail that those lines are perpendicular to each other (the relation is not transitive).

By its nature, the equivalence relation R on a set A leads to the concept of "equivalence class" which is defined as the set of elements in A that is related by R to a given element a in that set or, equivalently, by denoting this set as $\bar{a}$, we could define it symbolically as follows:

$$\bar{a} = \{x | x \in A \wedge xRa\} \tag{2.14}$$

From this definition and the properties of the equivalence class relation, for any $a \in A$ we have aRa (since R is reflexive) and hence $a \in \bar{a} \neq \phi$ (from the definition of $\bar{a}$). On the other hand, if $x \in \bar{a}$ and $x \in \bar{b}$ then $a, b \in \bar{x}$ (since R is symmetric) and hence $\bar{a} = \bar{b}$. Otherwise $\bar{a} \neq \bar{b}$. Finally, from the first proposition which asserts that the preceding results are true for all $a \in A$, we conclude that A is the union of all the equivalence classes created by the relation R. In general, we call the set which consists of the subsets of a set A, that has the previous properties, a "partition" of the set A. For instance, if R is a relation defined on $\mathbb{Z}$ (the set of integers) as follows:

$$(x, y) \in R \Rightarrow \exists\ q \in \mathbb{Z} \text{ such that } x - y = qn \tag{2.15}$$

then for any x, y and $z \in \mathbb{Z}$, we would have:

$$x - x = 0 \times n \text{ and } x - y = qn \Rightarrow y - x = -qn \tag{2.16}$$

also we have:

$$(x-y = q_1 n \wedge y-z = q_2 n) \Rightarrow x-z = (q_1+q_2)n = qn \quad (2.17)$$

so this relation is reflexive, symmetric and transitive and hence it constitutes an equivalence relation on $\mathbb{Z}$. Therefore it creates a partition (the set of equivalence classes) of the set of integers. We will refer to this partition as the set of residue classes modulo n and denote it as $\bar{\mathbb{Z}}_n$. Suppose that R is an equivalence relation, then for any x and y related to a given element a by R, the reflexive property of that relation would produce $xRa \Rightarrow aRx$ and hence, by the transitive property, we have:

$$yRa \text{ and } aRx \Rightarrow yRx \Rightarrow \bar{x} = \bar{a} \quad (2.18)$$

therefore any element in a given equivalence class can be used to represent that class. For instance, the equality relation defined on the set of all vectors in the plane is an equivalence relation which partitions that set into equivalence classes. Each of them represents a vector in the plane, so any vector that is an element of a given equivalence class of those can be used to represent that class. However, for simplicity, the vectors, that have the origin as the coordinates of their tails, are usually used to represent those classes.

A relation which does not satisfy the symmetry condition, except for the same element, is called "anti-symmetric" relation. If a given relation R on a set A is reflexive, anti-symmetric and transitive, we call it a "partial order" relation and if it, also, does not have the reflexivity property, we call it a "strict partial order" relation. However, if that relation, furthermore, has the following property:

$$\forall\ x, y \in A : (x, y) \in R \vee (y, x) \in R \quad (2.19)$$

we call it a "total order" and "strict total order" relation on A respectively. For instance, "the inclusion or equal" relation on the power set of a given set A is a partial order relation on that set and "the less than or equal" relation on the set of integers $\mathbb{Z}$ is a total order relation. On the other hand, the "inclusion" and "less than" relations on these sets, respectively, constitute a strict partial order and strict total order relations.

On many occasions, we are interested in a particular pattern of relations in which the domain coincides with the departure and any element in the domain is related by that relation with only one element in the range. We call this type of relation a "mapping". More precisely, a relation is called a mapping if it satisfies the following two conditions:

$$D = \{x | \ x \in A \wedge (x, y) \in R\} = A \tag{2.20}$$

and

$$(x, y) \wedge (x, z) \in R \Rightarrow y = z \tag{2.21}$$

if f is a mapping from a set A to another set B (denoted as $f : A \longrightarrow B$), we call $y \in A$, such that $(x, y) \in f$, the image of x by f and denote that as $y = f(x)$. More generally, if $A^* \subseteq A$, we define its image as follows:

$$f(A^*) = \{y \in B | \ y = f(x) \wedge x \in A\} \tag{2.22}$$

if f is a mapping such that $f(x_1) = f(x_2) \Rightarrow x_1 = x_2$, we call it an "injective" or "one to one" mapping, and if the image of its entire domain is its destination (that is $f(A) = B$), we call it a "surjective". However, if a mapping is both injective and surjective, it is called a "bijective" mapping. For example, the mapping $f : \mathbb{Z} \longrightarrow \mathbb{Z}$ defined by $f(x) = x + 5$ is a bijective, since any element in its domain (the set of integer numbers $\mathbb{Z}$) is related to only one element in its range (again the set $\mathbb{Z}$) and the image of its entire domain is the destination (that is $f(\mathbb{Z}) = \mathbb{Z}$), so it is injective and surjective which entails that it is a bijective mapping by definition. On the other hand, the mapping $f(x) = x^2 + 1$ defined on the same set (the set of integer numbers $\mathbb{Z}$) is not one to one since, for example, $f(-2) = f(2) = 4 + 1 = 5$ and $-2, 2 \in \mathbb{Z}$, also it is not surjective because all of the negative elements along with the zero and many positive elements, in the destination, have no inverse-images, by this mapping, in the domain. Therefore this mapping is not bijective. As an illustration of those three types of mapping, let's define the following three relations f_1, f_2 and f_3 from A_1 to A_2 (the previously defined two sets):

$$\begin{aligned} f_1 &= \{(a,d),(b,e)\} \\ f_2 &= \{(a,f,),(b,e),(c,e)\} \\ f_3 &= \{(a,e),(b,d),(c,f)\} \end{aligned}$$

it is obvious, from the relative definitions, that f_1, f_2 and f_3 is injective, surjective and bijective mappings, respectively. The arrow diagrams in Figure 2.2 constitute visual intermediate aspects of these mappings.

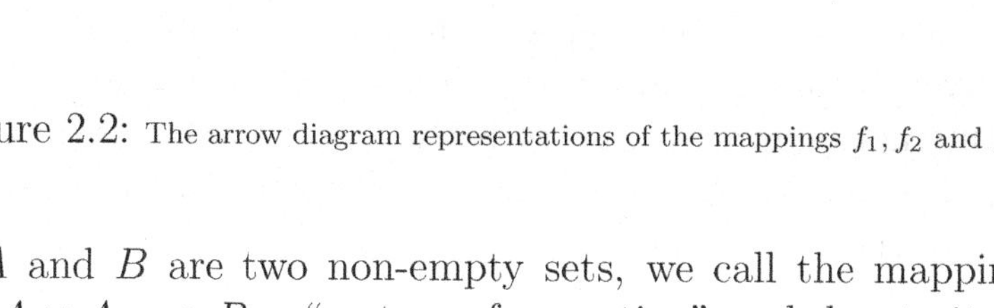

Figure 2.2: The arrow diagram representations of the mappings f_1, f_2 and f_3.

If A and B are two non-empty sets, we call the mapping $f : A \times A \longrightarrow B$ a "system of operation" and denote it as $(A, \star)$. If $B \subseteq A$ then any two elements of A are related, by that mapping, to an element which is again in A hence the system is described as a "closed system" or "magma" and the operation is called a "binary operation" on A. However, if $B \not\subseteq A$, then there would exist some pairs of elements from A that are related to some elements which are not in A, so the system, in this case, is designated as a "not closed system". As an example, the system $(\mathbb{R}, \star)$ defined as:

$$\forall \ (x, y) \in \mathbb{R} \times \mathbb{R} : \ x \star y = x + y \tag{2.23}$$

constitutes a closed system (magma) and hence its operation is a binary operation, since the addition of any two real numbers is a unique real number. For the same line of thought, the ordinary multiplication operation defined on $\mathbb{R}$ is also a binary operation. On the other hand, the ordinary subtraction operation defined on the positive real numbers $\mathbb{R}^+$ is not a binary relation, since $x - y \leq 0 \notin \mathbb{R}^+$ for all $x, y \in \mathbb{R}^+$ such that $x \leq y$ hence the system $(\mathbb{R}^+, -)$ is not a closed system. We call an operation $\star$ on a set a commutative and associative if it satisfies the respective conditions:

$$\forall \, x, y, z \in A : x \star y = y \star x \ \text{ and } \ (x \star y) \star z = x \star (y \star z) \tag{2.24}$$

if there is an element e in the set A such that $x \star e = x$ or $e \star x = x$, we call it the right or the left neutral element, respectively. However, if both conditions are satisfied, it is called the "two-sided neutral element" or simply a "neutral element". We notice that if e_1 and e_2 are two natural elements of a given binary operation $\star$ then, from the definition, we have $e_1 \star e_2 = e_1 = e_2$, so if the natural element exists for a given binary operation, it will be a unique element. A relating notion to this concept is the notion of the "inverse" of an element. We call an element a in the set A that satisfies one of the conditions:

$$x \star a = e \ \text{ or } \ a \star x = e \tag{2.25}$$

a right inverse or a left inverse of x, respectively. If both of these stipulations are true for a, it is called a "two-sided inverse" of x or in short an inverse of x and we denote it as x^{-1}. However, if $\star$ is an associative binary operation and x_1 and x_2 are two inverses of the element x then, from the related definitions, we have $(x_1 \star x) \star x_2 = x_2 = x_1 \star (x \star x_2) = x_1$ hence the associative property of a binary operation is sufficient for the inverse, if exist, to be unique.

The union operation "$\cup$" on the set that consists of all possible sets constitutes an example of a binary operation that is commutative, associative and has a natural element, but the inverse does not exist for any one of its elements. The composition operation "$\circ$" of mappings on the set of all mappings is a binary operation that is associative, has a natural element (the identity mapping I) and an inverse for each element, but it is not commutative. However, the addition operation "$\oplus$" defined on the set of residue classes modulo n ($\bar{\mathbb{Z}}_n$) is commutative, associative, has a natural element and an inverse for each element.

For many applications, it is more convenient to consider two operations $\star$ and $\circ$ defined on a given set A. If each of these operations forms a closed system with respect to this set, we call them, collectively, a "system of two operations" and refer to them, symbolically, as $(A, \star, \circ)$. The "distributive property" is a mutual property of these two operations that is of a particular interest. For every x, y and z in A, the distribution of the operation $\circ$ over the operation $\star$ from left and right can be defined, respectively, as follows:

$$x \circ (y \star z) = (x \circ y) \star (x \circ z) \text{ and } (y \star z) \circ x = (y \circ x) \star (z \circ x) \quad (2.26)$$

a system of two operations is designated as having the property of the distribution if both of these stipulations are satisfied. For instance, the system $(\mathbb{C}, +, \times)$, where the operations $+$ and $\times$ are the addition and multiplication defined on the set of complex numbers $\mathbb{C}$, has the distributive property.

This follows, from the related definitions, as stated below:

$$\begin{aligned} z_1 \times (z_2 + z_3) &= z_1 \times z_2 + z_1 \times z_3 \\ &= z_2 \times z_1 + z_3 \times z_1 = (z_2 + z_3) \times z_1 \end{aligned}$$

A given set A, as an orthodox meaning, has an infinitely many possible intermediate aspects (names). Some of them are very compact and others are elaborated (with many compactness levels between them). For instance, the pronunciation of the letter "A" can be used as an audible compact middle aspect of this set and its drawing as a visual compact intermediate aspect of it. Also, another set B could be utilized as an elaborated (audible or visual) middle aspect of the set A. In the latter case, some or all of the elements in set B stand as intermediate aspects (names) of the elements in set A, and there is a number of possible matchings that depends on the number of the elements in each set. However, we are interested in the particular matching f in which every element in A has one middle aspect (name) in the set B which we have called a mapping from the set A to the set B. On the other hand, as an orthodox meaning, any binary operation $\star$ that is defined on A, would have an intermediate aspect (name) that constitutes an operation $\circ$ defined on the set B. To create this "name" of the operation $\star$ on A, we define the operation $\circ$ on B such that for every $a_1, a_2 \in A$ we have:

$$f(a_1 \star a_2) = f(a_1) \circ f(a_2) \tag{2.27}$$

we call the mapping f, that satisfies the previous stipulation, a "homomorphism". If the homomorphism is injective, surjective and bijective, it is called "monomorphism", "epimorphism" and "isomorphism", respectively and if $A = B$ the homomorphism and the isomorphism are, respectively, called "endomorphsim" and "automorphism". Therefore, in general, if a homomorphism from the system $(A, \star)$ to the system $(B, \circ)$ did exist, the operation $\circ$ would be a "name" (intermediate aspect) of the operation $\star$. For example, the systems $(\mathbb{R},+)$ and $(\mathbb{R},\times)$, which constitute the ordinary addition and multiplication on the set of real numbers, have an isomorphism between them and hence the operation $+$ is a name (middle aspect) of the operation $\times$. This fact is demonstrated below by using the function $f(a) = \ln a$ as one form of the homomorphism that exists between those systems:

$$f(a_1 \times a_2) = \ln(a_1 \times a_2) = \ln a_1 + \ln a_2 = f(a_1) + f(a_2) \quad (2.28)$$

The preceding arguments can be generalized to the systems $(A,\star,\circ)$ and $(B,\oplus,\otimes)$ of two operations by requiring the existence of a mapping f with the following properties (a homomorphism) between them:

$$\begin{aligned} \forall\, a_1, a_2 \in A : f(a_1 \star a_2) &= f(a_1) \oplus f(a_2) \quad (2.29) \\ f(a_1 \circ a_2) &= f(a_1) \otimes f(a_2) \end{aligned}$$

As an illustration of this phenomenon, this type of mapping does exist between the two systems $(\mathbb{R},+,\times)$ and $(\mathbb{R}^+,\boxplus,\boxtimes)$, where the operations in the first system are the common addition and multiplication defined on the set of real numbers and the operations in the second system are defined on the set of the positive real numbers as follows:

$$a \boxtimes b = a^{\ln b} \quad \text{and} \quad a \boxplus b = a \times b \quad (2.30)$$

so if we define a mapping $f : \mathbb{R}^+ \longrightarrow \mathbb{R}$ such that $f(a) = \ln a$ we would have:

$$\begin{aligned} f(a \boxtimes b) &= \ln(a^{\ln b}) = \ln a \times \ln b = f(a) \times f(b) \quad (2.31) \\ f(a \boxplus b) &= \ln(a \times b) = \ln a + \ln b = f(a) + f(b) \end{aligned}$$

so the mapping f is a homomorphism from $\mathbb{R}^+$ to $\mathbb{R}$ and hence the operations $\boxplus$ and $\boxtimes$ are names (intermediate aspects) of the operations $+$ and $\times$, respectively. Suppose that $(A, \star)$ and $(B, \circ)$ are closed systems and $f : A \longrightarrow B$ is a homomorphism between them. Then the system $(f(A), \circ)$, where $f(A) \subseteq B$, is a closed system and its algebraic structure is inherited from algebraic structure of the system $(A, \star)$. This can be deduced, from the stated conditions, as follows:

$$f(a_1 \star a_2) = f(a_1) \circ f(a_2) = b_1 \circ b_2 \Rightarrow (b_1 \circ b_2) \in f(A)$$

and

$$\begin{aligned} f(a_1 \star a_2) = f(a_1) \circ f(a_2) &= b_1 \circ b_2 \\ &= f(a_2 \star a_1) \\ &= f(a_2) \circ f(a_1) = b_2 \circ b_1 \end{aligned}$$

also we have

$$\begin{aligned} f(a_1 \star (a_2 \star a_3)) &= f(a_1) \circ f(a_2 \star a_3) \\ &= f(a_1) \circ (f(a_2) \circ f(a_3)) \\ &= b_1 \circ (b_2 \circ b_3) \\ &= f((a_1 \star a_2) \star a_3) \\ &= f(a_1 \star a_2) \circ f(a_3) \\ &= (f(a_1) \circ f(a_2)) \circ f(a_3) = (b_1 \circ b_2) \circ b_3 \end{aligned}$$

finally

$$f(a_1 \star e) = f(a_1) \circ f(e) = f(a_1)$$

and

$$f(a_1 \star a_1^{-1}) = f(a_1) \circ f(a_1^{-1}) = f(e)$$

where $a_1, a_2, a_3 \in A$ and $f(a_1) = b_1, f(a_2) = b_2, f(a_3) = b_3 \in f(A)$. Therefore, any algebraic structure of the system $(A, \star)$ will be inherited by the system $(f(A), \circ)$. This entails that if the mapping $f : A \longrightarrow B$ is a bijection (f is an isomorphism) then any one of the systems $(A, \star)$ and $(B, \circ)$ can be used as a "name" (intermediate aspect) of the other. (We call them "isomorphic" and denote that by $A \cong B$.) In fact, if A and B are two sets with the same number of elements and $\star$ is an operation defined on A, we can always create an operation $\circ$ on B based on any bijection f between them such that:

$$f(a_1 \star a_2) = f(a_1) \circ f(a_2) \tag{2.32}$$

therefore all the previous results will hold and hence each of these systems can be employed as a middle aspect or name of the other. Moreover, for any given two systems $(A, \star)$ and $(B, \circ)$, with the same properties and number of elements, we can always define a bijection f between them so that the previous condition is satisfied and hence the two systems are isomorphic and $A \cong B$. Of course, it is always possible to use any system $(B, \circ)$ as a "name" of another system $(A, \star)$, even if they do not have the same number of elements, and create the operation $\circ$ on B based on any matching between them or create a matching base on any operation $\circ$ defined on B. It is also possible that the matching f does not constitute a mapping between them. However, as we depart from the bijection and the associated isomorphism (the ideal conditions), we lose more and more properties which are very useful in many situations. We call the representation of relations, based on the ideal conditions, the "orthodox representation" and the relations represented in that way, the "orthodox relations", so the orthodox names (intermediate aspects) are the most convenient "names" of relations.

If the operation $\star$ in a given system $(A, \star)$ is associative, we designate the system as an associative system and, for brevity, refer to it as a "semi-group". If a system is a semi-group and has a neutral element and inverse for each element, we call it a "group". Furthermore, if the operation $\star$ is commutative, the system is referred to as a commutative system and if it constitutes a group with respect to the operation $\star$, it is called a "commutative group". For example, the system $(\mathbb{C}, \times)$, where $\times$ is the common multiplication operation defined on the set of complex numbers, is a commutative group. For any three groups $(A_1, \star), (A_2, \bullet)$ and $(A_3, *)$, if f and g are two homomorphisms such that $f : A_1 \longrightarrow A_2$ and $g : A_2 \longrightarrow A_3$, then for every $a_1, a_2 \in A_1$ and h, where $h = g \circ f$, we have:

$$\begin{aligned} h(a_1 \star a_2) &= (g \circ f)(a_1 \star a_2) \\ &= g(f(a_1 \star a_2)) \\ &= g(f(a_1) \bullet f(a_2)) \\ &= g(f(a_1)) * g(f(a_2)) \\ &= (g \circ f)(a_1) * (g \circ f)(a_2) = h(a_1) * h(a_2) \end{aligned}$$

so the composition $h = g \circ f$ is a homomorphism from A_1 to A_3 and, if both f and g are bijective (that is both of them are isomorphisms), then h is a bijective (an isomorphism) and $A_1 \cong A_2 \cong A_3$. This result can be generalized, using the mathematical induction, to any number of groups.

If a set G, with finite number of elements n, constitutes a group with respect to an operation $\star$, we call G a "finite group" of order $|G| = n$. Otherwise, it is called an "infinite group". The set $G = \{1, i, -1, -i\}$, with respect to the common multiplication $\times$, represents an example of a finite group of order 4. This result is evident in the following multiplication table of this operation:

×	1	i	-1	-i
1	1	i	-1	-i
i	i	-1	-i	1
-1	-1	-i	1	i
-i	-i	1	i	-1

Table 2.1: The multiplication operation $\times$ defined on the set G.

If A and B constitute two groups with respect to an operation $\star$ and $B \subseteq A$, then the group B is called a "subgroup" of the group A and is denoted by $B \leq A$. For instance, the preceding group $G = \{1, i, -1, -i\}$ is a subgroup of the group of complex numbers $(\mathbb{C}, \times)$. It is sufficient for a set $B \subseteq A$ to be a subgroup of the group $(A, \star)$, if $B \neq \phi$ and for every $b_1, b_2 \in B$ we have $b_1 \star b_2^{-1} \in B$. The following argument shows how these conditions entail the conclusion that $B \leq A$:

$$b_1, b_1 \in B \Rightarrow b_1 \star b_1^{-1} \in B \Rightarrow e \in B \tag{2.33}$$

and hence

$$e, b_1 \in B \Rightarrow e \star b_1^{-1} \in B \Rightarrow b_1^{-1} \in B \tag{2.34}$$

so we have

$$b_1, b_2 \in B \Rightarrow b_1, b_2^{-1} \in B \Rightarrow b_1 \star b_2 \in B \tag{2.35}$$

Therefore, this demonstrates that the previous conditions necessitate the existence of a neutral element and an inverse for each elements in B. Also it shows that B is a closed system, so since, as a subset of A, B inherits the associative property of A, we conclude that B is a subgroup of the group A.

If $(A, \star)$ is a closed system and $a \in A$, then $a \star a \star ... \star a \in A$, and if this operation is applied n times on the element a, it is denoted as a^n. This notation can be generalized to negative integer powers of a by letting $a^{-1} \star a^{-1} \star ... \star a^{-1} = a^{-n} \in A$, where the operation $\star$ is iterated n times. As a special case, we use the definition $a^0 = e$, where e is the neutral element of the operation $\star$. If we have a group $(A, \star)$ and $a \in A$, then the set of all possible integer powers of a constitutes a group with respect to the operation $\star$, which we denote as $(\langle a \rangle, \star)$, and since $(a \star a \star ... \star a), (a^{-1} \star a^{-1} \star ... \star a^{-1}) \in A$, this group is a subgroup of A. This result can be proven as follows:

$$\langle a \rangle = \{a^n | n \in \mathbb{Z}\} \subseteq A \Rightarrow a^1 = a \in \langle a \rangle \Rightarrow \langle a \rangle \neq \phi \quad (2.36)$$

and for every $n_1, n_2 \in \mathbb{Z}$ we have

$$n_2 \in \mathbb{Z} \Rightarrow -n_2 \in \mathbb{Z} \Rightarrow (n_1 - n_2) \in \mathbb{Z} \quad (2.37)$$

hence

$$a^{n_1}, a^{n_2} \in \langle a \rangle \Rightarrow a^{n_1} \star a^{-n_2} = a^{(n_1 - n_2)} \in \langle a \rangle \quad (2.38)$$

from this argument, we conclude that $\langle a \rangle \leq A$. Naturally, we call the element $a \in \langle a \rangle$ the "generator" of the group $(\langle a \rangle .\star)$ and if a given group $(A, \star)$ has a generator, we call it a "cyclic group". It is obvious that any group $\langle a \rangle$ generated by the element a is a cyclic group. As an example, the system $(\mathbb{Z}_5, \oplus)$, where the operation $\oplus$ is defined as $a \oplus b = r$ such that there exist two numbers $q, r \in \mathbb{Z}$ and $0 \leq r \leq |b|$ for which $a = qb + r$, is a cyclic group since:

$$\langle 1 \rangle = \{1, 1^2, 1^3, 1^4, 1^5\} = \{1, 2, 3, 4, 0\} = \mathbb{Z}_5 \quad (2.39)$$

Let A be defined as $A = \{a | \ n \in \mathbb{N} \wedge 1 \leq a \leq n\}$, where $\mathbb{N}$ is the set of natural numbers. We call the mapping $f : A \longrightarrow A$ a permutation, since it just permutes the elements of A. It is obvious that this mapping is a bijection because the image of the entire departure A is the entire destination which is again A. The set of all possible permutations of the set A (denoted as A_n) constitutes a group with respect to the composition of mappings $\circ$. This results from the definition of this set since we have:

$$\forall \ f, g \in A_n \Rightarrow f \circ g \in A_n \quad (2.40)$$

however, for each $f \in A_n$ there is $f^{-1} \in A_n$ such that $f \circ$

$f^{-1} = I \in A_n$, and since the operation of composition of mappings is associative, it follows that A_n forms a group with respect to this operation which is called a "symmetric group" of degree n. For convenience, we represent a mapping $f \in A_n$ as below:

$$f = \begin{pmatrix} 1 & 2 & \ldots & n \\ f(1) & f(2) & \ldots & f(n) \end{pmatrix}$$

To appreciate the connection between these types of groups and "symmetry", we will consider the symmetric group $S_4 \leq A_4$ defined as follows:

$$S_4 = \left\{ \begin{pmatrix} 1\,2\,3\,4 \\ 1\,2\,3\,4 \end{pmatrix}, \begin{pmatrix} 1\,2\,3\,4 \\ 3\,4\,1\,2 \end{pmatrix}, \begin{pmatrix} 1\,2\,3\,4 \\ 2\,1\,4\,3 \end{pmatrix}, \begin{pmatrix} 1\,2\,3\,4 \\ 4\,3\,2\,1 \end{pmatrix} \right\}$$

if we let $H = \{r_1, r_2, f_1, f_2\}$ be the set that constitutes the symmetries of a rectangle with vertexes a, b, c, d, where these geometrical transformations (two rotations about its center by $r_1 = 360°$ and $r_2 = 180°$ and two reflections about its axes of symmetry) are mappings defined on the plane as below:

$$r_1 = \begin{pmatrix} a\;b\;c\;d \\ a\;b\;c\;d \end{pmatrix}, r_2 = \begin{pmatrix} a\;b\;c\;d \\ c\;d\;a\;b \end{pmatrix}$$

and

$$f_1 = \begin{pmatrix} a\;b\;c\;d \\ b\;a\;d\;c \end{pmatrix}, f_2 = \begin{pmatrix} a\;b\;c\;d \\ d\;c\;b\;a \end{pmatrix}$$

then the composition of mappings $\circ$ defined as an operation on the set of the symmetries of this rectangle forms a commutative group on that set. This fact can be demonstrated by constructing a table for that operation (see table 2.2). However, since the symmetric group S_4 has the same order as H and forms a commutative group with respect to the composition operation, then $H \cong S_3$ and each one of them can be employed as an orthodox name for the other.

$\circ$	r_1	r_2	f_1	f_2
r_1	r_1	r_2	f_1	f_2
r_2	r_2	r_1	f_2	f_1
f_1	f_1	f_2	r_1	r_2
f_2	f_2	f_1	r_2	r_1

Table 2.2: The composition operation $\circ$ defined on the set H.

For any subgroup $(B, \star)$ of a group $(A, \star)$, we call the two sets $a \star B = \{a \star b|\ b \in B\}$ and $B \star a = \{b \star a|\ b \in B\}$, where $a \in A$, the left and right "cosets" of the subgroup B, respectively, and the element a is called the representative of these sets. For instance, if A_3 is the symmetric group of the third degree and B is the subgroup of it defined by:

$$B = \left\{ \begin{pmatrix} 1\,2\,3 \\ 1\,2\,3 \end{pmatrix}, \begin{pmatrix} 1\,2\,3 \\ 2\,3\,1 \end{pmatrix}, \begin{pmatrix} 1\,2\,3 \\ 3\,1\,2 \end{pmatrix} \right\}$$

then, for the representative x, defined below, we have:

$$x = \begin{pmatrix} 1\,2\,3 \\ 3\,2\,1 \end{pmatrix} \Rightarrow x \circ B = \left\{ \begin{pmatrix} 1\,2\,3 \\ 3\,2\,1 \end{pmatrix}, \begin{pmatrix} 1\,2\,3 \\ 1\,3\,2 \end{pmatrix}, \begin{pmatrix} 1\,2\,3 \\ 2\,1\,3 \end{pmatrix} \right\}$$

and

$$B \circ x = \left\{ \begin{pmatrix} 1\,2\,3 \\ 3\,2\,1 \end{pmatrix}, \begin{pmatrix} 1\,2\,3 \\ 2\,1\,3 \end{pmatrix}, \begin{pmatrix} 1\,2\,3 \\ 1\,3\,2 \end{pmatrix} \right\}$$

hence the sets $x \circ B = B \circ x$ constitute, respectively, left and right cosets of the subgroup B. It is not necessary that the left coset to be identical to the right coset for a given subgroup (as in the previous example). However, if both of them are equivalent, we call the related subgroup a "normal subgroup" and refer to that by the notation $B \trianglelefteq A$. If B

is a subgroup of a finite group A with the respective orders $|B|= m$ and $|A|= n$, and if we define a relation R on A as follows:

$$\forall\, a_1, a_2 \in A : (a_1, a_2) \in R = a_1 \star a_2^{-1} \in B \tag{2.41}$$

then for every $a_1, a_2, a_3 \in A$ we have:

$$e \in B \Rightarrow a_1 \star a_1^{-1} \in B \Rightarrow (a_1, a_2) \in R$$

and

$$\begin{aligned} (a_1, a_2) \in R = a_1 \star a_2^{-1} \in B &= (a_1 \star a_2^{-1})^{-1} \in B \\ &= a_2 \star a_1^{-1} \in B = (a_2, a_1) \in R \end{aligned}$$

also we have

$$\begin{aligned} (a_1, a_2) \;\wedge\; (a_2, a_3) \in R &\Rightarrow a_1 \star a_2^{-1} \in B \;\wedge a_2 \star a_3^{-1} \in B \\ &\Rightarrow a_1 \star a_2^{-1} \star a_2 \star a_3^{-1} \in B \\ &\Rightarrow a_1 \star a_3 \in B = (a_1, a_3) \in R \end{aligned}$$

so R is an equivalence relation that partitions the set A into the set $P = A/R$ of equivalence classes, and since A is finite, P has a finite order $|P|= r$. However, from the related definitions, for every equivalence class $\bar{a_1}$ in P we have:

$$\begin{aligned} a \in \bar{a_1} = (a, a_1) \in R &= [(a \star a_1^{-1}) = b] \in B \\ &= a \in B \star a_1 \Rightarrow \bar{a_1} = B \star a_1 \end{aligned}$$

so the set P of the equivalence classes is, in fact, the set of all the right cosets of the set B and since the equivalence classes constitute a partition of the set A, we have:

$$\begin{aligned} n = |A| &= |\bar{a_1}|+|\bar{a_2}|+ ... + |\bar{a_r}| \\ &= |B \star a_1|+|B \star a_2|+ ... + |B \star a_r| \\ &= m + m + ... + m = rm = r|B| \end{aligned}$$

therefore, the set of the equivalence classes P has the order $|P|= r = [A : B]$. On the other hand, suppose that A and B are any two groups, where $B \trianglelefteq A$. If we define an operation $\bullet$ on the set $H = \{a \star B | a \in A\}$ such that for every $a_1, a_2 \in A$, we have:

$$(a_1 \star B) \bullet (a_2 \star B) = (a_1 \star a_2) \star B \tag{2.42}$$

then the left coset of B constitutes a group of order $[A : B]$. This result can be demonstrated as follows:

$$\forall\, a_1, a_2 \in A \;\exists\; b_1, b_2 \in B : \;\; a_1 = a_1' \star b_1 \;\; \text{and} \;\; a_2 = a_2' \star b_2$$

because $\star$ is a binary operation, and since B forms a group with respect to this operation, we have:

$$\begin{aligned}(a_1 \star a_2) \star B &= (a_1' \star b_1 \star a_2' \star b_2) \star B\\ &= (a_1' \star b_1 \star a_2') \star B\\ &= (a_1' \star a_2' \star b_3) \star B = (a_1' \star a_2') \star B\end{aligned}$$

where $a_2' \star b_3 = b_1 \star a_2'$ since B is a normal subgroup of A. This proves that the system $(H, \bullet)$ is a closed system. However, for every $a_1 \star B = x, a_2 \star B = y, a_3 \star B = z \in H$, we have:

$$\begin{aligned}(x \bullet y) \bullet z &= [(a_1 \star B) \bullet (a_2 \star B)] \bullet (a_3 \star B)\\ &= [(a_1 \star a_2) \star B] \bullet (a_3 \star B)\\ &= [(a_1 \star a_2) \star a_3] \star B\\ &= [a_1 \star (a_2 \star a_3)] \star B\\ &= (a_1 \star B) \bullet [(a_2 \star a_3) \star B]\\ &= (a_1 \star B) \bullet [(a_2 \star B) \bullet (a_3 \star B)] = x \bullet (y \bullet z)\end{aligned}$$

so the system is associative, and since for every $(b \star B) = x \in H$, there exist two elements $(e \star B) = e^*, (b \star B)^{-1} = (b^{-1} \star B) = x^{-1} \in H$ such that:

$$e^* \bullet x = (e \star B) \bullet (b \star B) \;=\; (e \star b) \star B$$

$$\begin{aligned} &= b \star B = (b \star e) \star B \\ &= (b \star B) \bullet (e \star B) \\ &= x \bullet e^* \end{aligned}$$

and

$$\begin{aligned} x^{-1} \bullet x = (b^{-1} \star B) \bullet (b \star B) &= (b^{-1} \star b) \star B \\ &= e \star B \\ &= e^* \\ &= (b \star b^{-1}) \star B \\ &= (b \star B) \bullet (b^{-1} \star B) \\ &= x \bullet x^{-1} \end{aligned}$$

so the system has a neutral element and an inverse for each of its elements which complete the proof, and hence the system $(H, \bullet)$ is a group, and since $B \trianglelefteq A \Rightarrow a \star B = B \star a$, it follows, from the definition of the operation $\bullet$ and the previous result, that $|H| = [A : B]$. The group H is called the "Quotient group" and is denoted by $H = A/B$. If $(A_1, \star)$ and $(A_2, \bullet)$ are two groups and $(B, \star)$ is another group such that $B \leq A_1$ and if $f : A_1 \longrightarrow A_2$ is a homomorphism between them, then for every $b_1, b_2 \in B \subseteq A_1$ there will be $a_1 = f(b_1), a_2 = f(b_2) \in A_2$ such that:

$$a_1 \bullet a_2^{-1} = f(b_1) \bullet f(b_2^{-1}) = f(b_1 \star b_2^{-1}) \in f(B)$$

so if we take $B = A_1$, as a consequence of the preceding argument, we have $f(A_1) \leq A_2$. However, if we suppose that $B \trianglelefteq A_1$ and notice that $b \star B = B \star b \Rightarrow b \star B \star b^{-1} = B$, then for every $b \in B$ and $b' = f(b) \in A_2$, we have:

$$a_1 \bullet b' \bullet a_1^{-1} = f(b_1) \bullet f(b) \bullet f(b_1^{-1}) = f(b_1 \star b \star b_1^{-1}) \in f(B)$$

and hence $f(B) \trianglelefteq f(A_1) \leq A_2$.

Any commutative group $(R, +)$ and semi-group $(R, \times)$

that have, as a common property, the distribution of the semi-group operation $\times$ on the commutative group operation $+$, are called, collectively, a "ring" and referred to them as a system of two operations by $(R, +, \times)$. The distributed operation is called, in general, a "multiplication" and the other operation is called an "addition". If the multiplication operation is commutative, we call the ring a "commutative ring" and if it has a neutral element, we call it a "unit ring". However, if the multiplication operation constitutes a commutative group, we call that system a "field" and denotes it by $(F, +, \times)$. For convenience, we will refer to the additive neutral element, multiplicative neutral element, additive inverse and multiplicative inverse as $0, 1, -x$ and x^{-1} respectively, and for brevity, the operation $x + (-y)$ is referred to simply as $x - y$. For example, the system $(\mathbb{Z}, +, \times)$, where $+$ and $\times$ are the common addition and multiplication, is a unit ring while the system $(\mathbb{R}, +, \times)$ is a field. Suppose that $(R, +, \times)$ is a ring. Then, from the related definitions, for every $r_1, r_2 \in R$ we have:

$$\begin{aligned}
0 = r_1 \times 0 + [-(r_1 \times 0)] &= r_1 \times (0 + 0) + [-(r_1 \times 0)] \\
&= r_1 \times 0 + [r_1 \times 0 - (r_1 \times 0)] \\
&= r_1 \times 0 + 0 = r_1 \times 0
\end{aligned}$$

and

$$\begin{aligned}
0 = 0 \times r_1 + [-(0 \times r_1)] &= (0 + 0) \times r_1 + [-(0 \times r_1)] \\
&= 0 \times r_1 + [0 \times r_1 - (0 \times r_1)] \\
&= 0 \times r_1 + 0 = 0 \times r_1
\end{aligned}$$

so we have

$$\begin{aligned}
0 = r_1 \times 0 &= r_1 \times (r_2 - r_2) \\
&= r_1 \times r_2 + r_1 \times -r_2 \Rightarrow r_1 \times -r_2 = -(r_1 \times r_2)
\end{aligned}$$

and

$$\begin{aligned} 0 = 0 \times r_1 &= (r_2 - r_2) \times r_1 \\ &= r_2 \times r_1 + (-r_2 \times r_1) \Rightarrow -r_2 \times r_1 = -(r_1 \times r_2) \end{aligned}$$

finally, from the previous results, if we let $r_3 = -r_1$, we would have

$$\begin{aligned} -r_1 \times -r_2 = r_3 \times -r_2 &= -(r_3 \times r_2) \\ &= -(-r_1 \times r_2) \\ &= -(-(r_1 \times r_2)) = r_1 \times r_2 \end{aligned}$$

therefore, if $(R, +, \times)$ is any ring, then for every $r_1, r_2 \in R$ we have: $r_1 \times 0 = 0 \times r_1 = 0, r_1 \times (-r_2) = (-r_1) \times r_2 = -(r_1 \times r_2)$ and $(-r_1) \times (-r_2) = r_1 \times r_2$. If the system $(R, +, \times)$ is a ring and $\phi \neq S \subseteq R$ constitutes a ring with respect to the same multiplication and addition operations defined on R, we call S a subring of the ring R and denote that by $S \leq R$. From the definition of the ring and the results of some previously proved groups theories, it is obvious that any "nonempty" subset S of a ring R is a subring of it, if it satisfies the following stipulations:

$$\forall\ s_1, s_2 \in S : s_1 - s_2 \in S \quad \text{and} \quad s_1 \times s_2 \in S \tag{2.43}$$

so a nonempty subset of a ring is a subring of it, if and only if, this subset is closed under its multiplication and subtraction. This proposition is called the "subring test". If R is a unit ring and $r \in R$ has a multiplicative inverse in R, we call r an "invertible element" or a "unit" in R. However, if $r_1, r_2 \in R$ such that $r_1 \times r_2 = 0$, we call r_1 and r_2 a left and right divisor of the zero, respectively, and we call R a "division ring" if the system $(R^*, \times)$, where $R^* = R - \{0\}$, is a group. As an illustration, for the ring $(\mathbb{Z}_{12}, +, \times)$ in which $a+b = a \oplus b = r$, where $\oplus$ is the operation defined, previously, on $\mathbb{Z}_5$, and the multiplication operation $\times$ is defined as:

$$\begin{aligned} a \times b = a_1 + a_2 + ... + a_b &= b_1 + b_2 + ... + b_a \\ &= a_1 \oplus a_2 \oplus ... \oplus a_b \\ &= b_1 \oplus b_2 \oplus ... \oplus b_a \end{aligned}$$

the elements $2, 3, 4$ and 6 are divisors of the zero in this ring, since $3\times 4 = 2\times 6 = 0$, and hence the equation $x^2-5x+6 = 0$ can be solved in this ring as follows:

$$\begin{aligned} x^2 - 5x + 6 = 0 &\Rightarrow (x-2) \times (x-3) = 0 \\ &\Rightarrow x - 2 = 2, 3, 4 \text{ or } 6 \\ &\Rightarrow x = 4, 5, 6 \text{ or } 8 \end{aligned}$$

and since only for 6 we have

$$\begin{aligned} x^2 - 5x + 6 = 0 &\Rightarrow (x-2) \times (x-3) = 0 \\ &\Rightarrow (6-2) \times (6-3) = 0 \\ &\Rightarrow 4 \times 3 = 0 \end{aligned}$$

but also we have

$$\begin{aligned} x^2 - 5x + 6 = 0 &\Rightarrow (x-2) \times (x-3) = 0 \\ &\Rightarrow x - 2 = 0 \vee x - 3 = 0 \\ &\Rightarrow x = 2 \vee x = 3 \end{aligned}$$

so the set $S = \{2, 3, 6\}$ constitutes the solution of this equation in $\mathbb{Z}_{12}$.

If a ring is commutative, has a unit and has no divisors to its zero, it is called a "perfect ring". However, the only stipulation for a given perfect ring R to be a field is to have a "finite" number of elements. To see why, let $R = \{1, r_1, r_2, ..., r_n\}$, where $|R|= n$, and let $r \neq 1 \in R$, then we will have:

$$(r_1 \times r = h_1), (r_2 \times r = h_2), ..., (r_n \times r = h_n)$$

so it follows that

$$h_1 \neq h_2 \neq ... \neq h_n$$

and hence there is at least one element in this sequence for which $h_i = 1$. This result will be true for any $r \neq 1 \in R$, so any element in R has a multiplicative inverse and the system $(R^*, \times)$ is a commutative group. Therefore the prescribed proposition holds.

We can generalize the previous discussion to systems of tripartite, quadripartite, etc. operations. However, the binary operations can be employed as basic blocks to describe any relations involving any number of elements. As an example, Coulomb's law can be viewed as a tripartite operation involving the three elements q_1, q_2 and r (which are, respectively, the two charges and the distance between them). So from this respect, it constitutes a mapping f from the third power of the set of real numbers to itself ($f : \mathbb{R}^3 \longrightarrow \mathbb{R}$), but we also have:

$$\begin{aligned} f(q_1, q_2, r) = \frac{q_1 \times q_2}{r^2} &= (q_1 \times q_2) \times (r^{-1} \times r^{-1}) \\ &= g(q_1, q_2) \times g(r^{-1}, r^{-1}) \\ &= r_1 \times r_2 = g(r_1, r_2) \end{aligned}$$

where $g(q_1, q_2) = r_1$ and $g(r^{-1}, r^{-1}) = r_2$. So the tripartite operation $f(q_1, q_2, r)$ has been reduced to the binary operation $g(r_1, r_2)$, and in general, if $\star$ and $\bullet$ are two binary operations on the same set A and f is a tripartite operation defined on A such that:

$$f(r_1, r_2, r_3) = (r_1 \star r_2) \bullet r_3 \quad \text{or} \quad f(r_1, r_2, r_3) = r_1 \star (r_2 \bullet r_3)$$

then we always have:

$$f(r_1, r_2, r_3) = (r_1 \star r_2) \bullet r_3 = r_4 \bullet r_3 = g(r_4, r_3)$$

and

$$f(r_1, r_2, r_3) = r_1 \star (r_2 \bullet r_3) = r_1 \star r_5 = h(r_1, r_5)$$

where $r_4 = r_1 \star r_2$ and $r_5 = r_2 \bullet r_3$. Therefore, by mathematical induction, this proposition can be generalized to operations with any number of operands.

2.2.4 Measurand systems

In many cases, it is convenient to treat a collection of concepts as a system, and in this case there should be some characteristics that isolate those concepts from the rest. One instance of this phenomenon, is the isolation of a collection of concepts by the "succession" relation. We define this relation on a given set A as follows:

$$\forall\ x, y \in A : xSy \Leftrightarrow y \text{ succeeds } x \tag{2.44}$$

however, it results from this definition, that for every $x, y, z \in A$ we have:

$$x \text{ does not succeed } x \Rightarrow x \not S x$$

and

$$y \text{ succeeds } x \Rightarrow x \text{ does not succeed } y \Rightarrow xSy \wedge y \not S x$$

and finally we have

$$\begin{aligned} y \text{ succeeds } x \ \wedge \ z \text{ succeeds } y &\Rightarrow z \text{ succeeds } x \\ &\Rightarrow xSy \wedge ySz \Rightarrow xSz \end{aligned}$$

so the succession is irreflexive, asymmetric and transitive and hence it constitutes a strict total order on the set A. The element $a_2 \in A$ is called the "inferior successor" of the element $a_1 \in A$ if a_2 succeeds a_1 and, if we also have:

$$\forall\ a_i \in A : a_i \text{ succeeds } a_1 \Rightarrow a_i \text{ succeeds } a_2 \tag{2.45}$$

we call the gap between an element and its inferior successor an "interval" (denoted by $[a_i \sim a_{i+1}]$) and refer to the pattern of all these intervals in the ordered set A as its "spacing". If all the intervals in a given set A, that isolated by a succession, are identical, we call the ordered set A a "uniform" ordered set. Otherwise it is called "nonuniform". If the ordered set A is nonuniform, we call its largest interval the "norm". However, an ordered set A is referred to as a "continuous" set if the following condition is satisfied:

$$\forall\ a, a' \in A\ \exists\ a'' \in A : aSa' \Rightarrow a''Sa'\ \wedge\ aSa'' \tag{2.46}$$

where S is a succession relation. If this stipulation is not satisfied, the ordered set A is called a "discrete" set. We, collectively, call the pattern of succession (continuous or discrete) a "sequence". An ordered set A has a "start point" a_0 and an "end point" a_n if the following condition is satisfied:

$$\forall\ a \in A : a_0 Sa \text{ and } aSa_n \tag{2.47}$$

A given ordered set A is called "finite" if it is discrete and has start and end points. Otherwise, it is called "infinite". An ordered set A is designated as a "closed" set if its start point coincides with its end point, and it is called an "open" ordered set if this condition is not satisfied. The ordinal aspect of the elements of a given ordered set A, which is created by the succession relation, could result from an inherent nature of the elements or it could be imposed by definition. For example, the set of natural numbers $\mathbb{N} = \{1, 2, 3, ...\}$

is ordered by nature. However, the set $A = \{a, b, c, ...\}$ of the alphabet letters is ordered by definition. We define the "measurand" as any element in a given set A isolated by a discrete or continuous succession ordered by nature or definition, and the ordered set A of these measurands is called a "measurand system". If A and B are two measurand systems, and if η is the bijection that satisfies the following condition: For every $a \in A$ and $b \in B$ there are $a_1, a_2 \in A$ and $\eta(a_1) = b_1, \eta(a_2) = b_2 \in B$ such that:

$$a_1 \text{ between } a \text{ and } a_2 \Rightarrow b_1 \text{ between } b \text{ and } b_2 \tag{2.48}$$

we call this mapping the "natural bijection". Any measurand system that is discrete and has a start point is called the "fundamental measurand system". For a given discrete measurand system with a given spacing, we call any fundamental measurand system with that spacing an "associated fundamental measurand system". The order of a given finite discrete measurand system (denoted by the vertical lines $|\ |$) is defined as the image of its end point, by the natural bijection, in an associated fundamental measurand system such that the image of its start point is the start point of the fundamental. Therefore, if F is an associative fundamental measurand system and f_0 is its start point, then the order of the related finite discrete measurand system A is defined as follows:

$$|A| = \eta(a_n) : \eta : A \longrightarrow F \quad \text{and} \quad \eta(a_0) = f_0 \tag{2.49}$$

so if we take the set of natural numbers $\mathbb{N}$, with a suitable spacing, as an associative fundamental measurand system, then the order of the related discrete finite measurand system will be the number of its elements. The order relation is an equivalence relation that divides the set of all measurand systems into equivalence classes. We call any one of these classes an "order" and refer to any element in a particular class by μ^w, where w is the order of that equivalence class using the set of whole numbers as a fundamental measurand system. If a measurand system consists of measurands which, themselves, constitute measurand systems, we call this system a "homogeneous" if all of its included measurand systems have the same order and spacing. Otherwise, it is designated as an "inhomogeneous" system. For instance, the decimal numeral system is a homogeneous measurand system and the D:M:S (degree, minute and second) system is an inhomogeneous measurand system.

The succession as a relation has the common visual intermediate aspects of relations. However, for convenience, we may use the "measurand line" and "measurand circle" as special middle aspects (names) of this relation for the respective case of the open and closed measurand systems. Figure 2.3 shows these visual intermediate aspects for a uniform discrete open measurand system of a finite order and a uniform discrete closed measurand system, respectively.

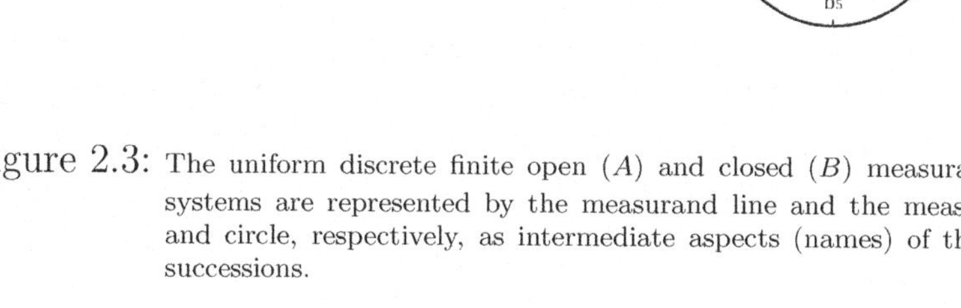

Figure 2.3: The uniform discrete finite open (A) and closed (B) measurand systems are represented by the measurand line and the measurand circle, respectively, as intermediate aspects (names) of their successions.

If a and a' are two measurands from a measurand system A, we write $a \prec a'$ or $a' \succ a$ when a precedes a' or a' succeeds a, respectively. The "auxiliary measurand system" and its inverse of a given measurand $a \in A$ (where A is a discrete measurand system) with respect to the reference ρ are defined, respectively, as follows:

$$\begin{aligned} \mathcal{A}_a^\rho &= \{x_1, x_2, ..., x_n\} \subseteq A : \\ &\qquad \rho = x_1 \prec x_2 \prec ... \prec x_n = a \end{aligned} \tag{2.50}$$

and

$$\begin{aligned} \mathcal{A}_{-a}^\rho &= -\mathcal{A}_a^\rho = \{-x_n, -x_{n-1}, ..., -x_1\} : \\ &\qquad \{x_1, x_2, ..., x_n\} \subseteq A \end{aligned} \tag{2.51}$$

where

$$-a = -x_n \prec -x_{n-1} \prec ... \prec -x_1 = \rho$$

the interval $[\rho = x_1 \sim x_2]$ between the reference ρ and the second element in an auxiliary measurand system is called the "fundamental interval". So any interval in a uniform measurand system is a fundamental interval, and since the order of the fundamental interval is "one" (using the set $\mathbb{W}$ as a fundamental measurand system), any element in its order will be denoted by μ. We define the operation $\overset{\rho}{+}$ on the discrete uniform measurand system A as below:

$$a \overset{\rho}{+} b = c \Leftrightarrow \eta_b(a) = c, \ \eta_b(\rho) = b \text{ and } \eta_b : \mathcal{A}_a^\rho \longrightarrow A \tag{2.52}$$

where η is the natural bijection. We call this operation a "mensuration" and since, by definition, we emphasized that the mensuration is an operation on A, then it constitutes a closed system on it. However, from the previous definition we have:

$$\begin{aligned} a \overset{\rho}{+} b = c_1 = \eta_b(a) \Rightarrow c_1 \in A_1 = \{a = x_1, x_2, \\ ..., x_n = b, \eta_b(y_1), \eta_b(y_2), \\ ..., \eta_b(y_n) = \eta_b(x_1) = \eta_b(a)\} \end{aligned}$$

but we also have

$$
\begin{aligned}
b \overset{\rho}{+} a = c_2 = \eta_a(b) \Rightarrow c_2 \in A_2 = \{a = x_1, x_2, \\
..., x_n = b, \eta_b(x_1), \eta_b(x_2), \\
..., \eta_b(x_n) = \eta_a(b)\} \Rightarrow c_2 \in \{a = \eta_a(y_1), \eta_a(y_2), \\
..., \eta_a(y_r) = b, \\
..., \eta_a(y_n) = \eta_a(a), \eta_b(x_1), \eta_a(x_2), ..., \eta_a(x_n) = \eta_a(b)\}
\end{aligned}
$$

therefore $|A_1| = |A_2|$, so since both A_1 and A_2 have the same start point and $A_1, A_2 \subseteq A$ then we conclude that $\eta_b(a) = c_1 = c_2 = \eta_a(b)$ and the operation $\overset{\rho}{+}$ is commutative. On the other hand, for every $a, b, c \in A$, we have, by appealing to the relevant definitions, the following results:

$$
\begin{aligned}
(a \overset{\rho}{+} b) \overset{\rho}{+} c = \eta_a(b) \overset{\rho}{+} c &= \eta_{\eta_a(b)}(c) \\
&= \eta_{\eta_b(a)}(c) \\
&= \eta_{\eta_b(c)}(a) \\
&= \eta_a(\eta_b(c)) \\
&= a \overset{\rho}{+} \eta_b(c) \\
&= a \overset{\rho}{+} (b \overset{\rho}{+} c)
\end{aligned}
$$

and

$$
\begin{aligned}
a \overset{\rho}{+} -a = \eta_{-a}(a) = c = \eta_a(-a) = c \\
\Rightarrow b \in \{\rho = x_1, x_2, ..., x_n = a\} \\
\text{and} \quad c \in \{-x_n = -a, -x_{n-1}, ..., -x_1 = \rho\}
\end{aligned}
$$

hence $c = \rho$. Also, by definition, we have:

$$\rho \overset{\rho}{+} a = \eta_a(\rho) = a$$

therefore, with respect to the measurand system $A^\star = -A \cup$

A, the operation $\overset{\rho}{+}$ is commutative, associative, has a neutral element and inverse for each of its elements and hence it constitutes a commutative group on the measurand system $A^{\star}$. As an illustration of this operation, if we take $\rho = 0$ and $A^{\star}$ to be the set of integer numbers $\mathbb{Z}$, the the operation $\overset{0}{+}$ will be the common addition $+$ defined on that set which constitutes a commutative group on it. Hence the addition is a special case of the mensuration. Suppose that $\mathfrak{D}$ is the set of all discrete finite measurand systems. We define the "nesting measurand system of frequency m" (denoted by $A_{\circlearrowleft}^{m}$) as follows:

$$\begin{aligned} A_{\circlearrowleft}^{m} = \{a_1 \prec a_2 \prec \ldots \prec a_n | \; \forall\, x \in A_{\circlearrowleft}^{m} \; \exists \; A_{\circlearrowleft}^{m-1} \in \mathfrak{D} : \\ x = \{x_1 \prec x_2 \prec \ldots \prec x_r | \; x_r \prec x\} = A_{\circlearrowleft}^{m-1}, \\ \ldots, \forall\, x \in A_{\circlearrowleft}^{2} \; \exists \; A \in \mathfrak{D} : \\ x = \{x_1 \prec x_2 \prec \ldots \prec x_s | \; x_s \prec x\} = A\} \end{aligned}$$

it results from this definition that any homogeneous and inhomogeneous measurand system is a nesting measurand system. We call the homogeneous nesting measurand system a "fractal measurand system" and if the frequency of the latter is infinity, we call it a "perfect fractal measurand system". On the other hand, when the frequency of a nesting measurand system approaches infinity, then, according to its definition, it will imply that for every $a, a' \in A_{\circlearrowleft}^{m}$ there is $a'' \in A_{\circlearrowleft}^{r}$, where $a' = A_{\circlearrowleft}^{r}$ such that:

$$aSa' \Rightarrow a''Sa' \;\wedge\; aSa''$$

hence any continuous measurand system is a union of nesting measurand systems of infinite frequencies. More precisely, if C is a continuous measurand system, then there is $n \in [0, \infty)$ such that:

$$C = \bigcup_{i=1}^{n} \lim_{m \to \infty} A_{\circlearrowleft i}^{m} \tag{2.53}$$

notice that, without loss of generality, we have used the whole numbers set $\mathbb{W}$ as a fundamental measurnd system to determine the order of i. However, if we assume that $A^k_{\circlearrowleft}$ is a finite nesting measurand system such that the binary system $(A^{k\star}_{\circlearrowleft}, \overset{\rho}{+})$ constitutes a commutative group, then the system $(A^{(k+1)\star}_{\circlearrowleft}, \overset{\rho}{+})$ is identical to the previous system except that it has one extra level of frequency. Therefore, this system constitutes a commutative group with respect to all the measurands above its final level (since they, collectively, represent the commutative group $(A^{k\star}_{\circlearrowleft}, \overset{\rho}{+})$ by assumption). Hence, to establish that the system $(A^{(k+1)\star}_{\circlearrowleft}, \overset{\rho}{+})$ also constitutes a commutative group, we just have to examine the final level of frequency. Suppose that $A^2 = x_1 \prec x_2 \prec ... \prec x_n$ and $A_{x_i} = y_1 \prec y_2 \prec ... \prec y_r$. So by definition, for any $y_i, y_{i+1} \in A_{x_i} = x_i$ we have $y_i \prec y_{i+1}$, but $x_{i-1} \prec y_1$ and $y_r \prec x_i$ for every $A_{x_i} = x_i$ and since $-A_{x_i} \subset A^{(k+1)\star}_{\circlearrowleft}$, we conclude that the system $(A^{(k+1)\star}_{\circlearrowleft}, \overset{\rho}{+})$ is a commutative group, and by mathematical induction, we know that this proposition is true for any value of $k \in \mathbb{W}$. Therefore, the system $(\lim_{k\to\infty} A^{k\star}_{\circlearrowleft}, \overset{\rho}{+})$ is a commutative group. This generalizes the operation $\overset{\rho}{+}$ to continuous measurand systems. As an example of a continuous measurand system that forms with the operation $\overset{\rho}{+}$ a commutative group, we may take the reference $\rho = 0$ and $\lim_{k\to\infty} A^{k\star}_{\circlearrowleft} = \mathbb{R}$ (the set of real numbers). This produces the commutative group $(\mathbb{R}, +)$ which is merely the common addition of real numbers.

Suppose that C is a continuous uniform measurand system and C_f is an associative fundamental measurand system, such that $[\rho \sim \mu]$ is the interval for the nesting level m, so that μ is the element of the order of the fundamental interval associated with the nesting level m (which is called the standard nesting level). We define the operation $\overset{\mu}{\times}$ on

$C^{\star} = -C \cup C$ as follows:

$$\begin{aligned} a \overset{\mu}{\times} b \;&=\; c \Leftrightarrow c = (a \overset{\rho}{+} a \overset{\rho}{+} \\ &\quad \ldots \overset{\rho}{+} a)_x : x = |b|, f_0 = \rho \\ &\qquad \text{and } [\rho \sim \mu] \in \mathfrak{I} \end{aligned} \tag{2.54}$$

where f_0 is the start point of the associative fundamental measurand system C_f, x is the order of the repetition and $\mathfrak{I}$ is the set of all intervals. We will refer to this operation as a "reduplication". The inverse of this operation (denoted by $\overset{\mu}{\div}$) is defined as follows:

$$a \overset{\mu}{\div} b = c \Leftrightarrow a = c \overset{\mu}{\times} b \tag{2.55}$$

From the definition of the operation $\overset{\mu}{\times}$ we have:

$$\begin{aligned} a \overset{\mu}{\times} b \;&=\; (a \overset{\rho}{+} a \overset{\rho}{+} \ldots \overset{\rho}{+} a)_x \\ &=\; (a \overset{\rho}{+} a \overset{\rho}{+} \ldots \overset{\rho}{+} a_r) \overset{\rho}{+} (a \overset{\rho}{+} a \overset{\rho}{+} \ldots \overset{\rho}{+} a_s) = c \overset{\rho}{+} d \end{aligned}$$

where $x = r \overset{\rho}{+} s$. So the reduplication inherits the associative and commutative properties of the operation $\overset{\rho}{+}$. On the other hand, since the reduplication of a measurand is, by definition, the repetition of that measurand by the order of the other operand, and since the order of the fundamental interval (the order of μ) just includes one element, then $a \overset{\mu}{\times} \mu = a$. So, from the commutative property, μ constitutes the neutral element of this operation. However, from the definition of the inverse of the reduplication and its commutative property, we have:

$$\forall\, a \in C^{\star} : (\mu \overset{\mu}{\div} a) \overset{\mu}{\times} a = a \overset{\mu}{\times} (\mu \overset{\mu}{\div} a) = \mu$$

hence, the element $c = (\mu \overset{\mu}{\div} a) \in C^{\star}$ (for brevity, we will use the notation $\frac{\mu}{a} = \mu \overset{\mu}{\div} a)$ is the inverse of the element $a \in C^\star$. Therefore, from the previous results, it is clear that the system $(C^\star, \overset{\mu}{\times})$ is a commutative group. If we take $C^\star = \mathbb{R}$ and $\mu = 1$, as a special case, then the resulting system $(\mathbb{R}, \times)$ constitutes the ordinary commutative group of multiplication. However, since for every $x, y, z \in C^\star$ we have:

$$\begin{aligned}
(x \overset{\rho}{+} y) \overset{\mu}{\times} z &= [(x \overset{\rho}{+} y) \overset{\rho}{+} (x \overset{\rho}{+} y) ... \overset{\rho}{+} (x \overset{\rho}{+} y)]_{|z|} \\
&= [(x \overset{\rho}{+} x \overset{\rho}{+} ... \overset{\rho}{+} x)_{|z|} \overset{\rho}{+} (y \overset{\rho}{+} y \overset{\rho}{+} ... \overset{\rho}{+} y)_{|z|}] \\
&= x \overset{\mu}{\times} z \overset{\rho}{+} y \overset{\mu}{\times} z \\
&= z \overset{\mu}{\times} x \overset{\rho}{+} z \overset{\mu}{\times} y = z \overset{\mu}{\times} (x \overset{\rho}{+} y)
\end{aligned}$$

and so the operation $\overset{\mu}{\times}$ distributes over the operation $\overset{\rho}{+}$ and hence the system $(C^\star, \overset{\rho}{+}, \overset{\mu}{\times})$ constitutes a field which we will refer to as the "orthodox field". The ordinary real field $(\mathbb{R}, +, \times)$ is an example of the orthodox field. From the previous results and the related definitions, the continuous nonuniform measurand system can be defined as follows:

$$C^* = \bigcup_{i=1}^{n} \lim_{m \to \infty} A^m_{\circlearrowleft i} \tag{2.56}$$

if and only if we have

$$\begin{aligned}
\forall\, a_k, a_{k+1} \in A^m_{\circlearrowleft r} \quad \exists \quad & b_k, b_{k+1} \in A^m_{\circlearrowleft s} : \\
& \mu_1 = [a_k \sim a_{k+1}] \neq [b_k \sim b_{k+1}] = \mu_2
\end{aligned}$$

from this definition, $\forall\ \eta_b : \mathfrak{D}^{\rho}_{a} \longrightarrow D$ and $\eta_b(\rho) = b\ \exists\ r \in [1, n]$ and $m^* < m$ such that:

$$\eta_b(a) = c \in A^m_{\circlearrowleft r} \quad \text{or} \quad \eta_b(a) = c \in A^{m^*}_{\circlearrowleft r}$$

so the mensuration operation, defined previously, is also applied to the continuous nonuniform measurand system C^*, and hence the reduplication operation is naturally extended to it by using its associative fundamental measurand system. Therefore, if $\mathcal{C} = C \cup C^*$ is any continuous measurand system which could be defined as follows:

$$\mathcal{C} = \bigcup_{i=1}^{n} \lim_{m \to \infty} A^m_{\mho i} : \quad \forall \quad a_k, a_{k+1} \in A^m_{\mho i} \qquad (2.57)$$
$$\exists\ \tau = [a_k \sim a_{k+1}] \in \mathcal{I}$$

then we define the "heteromensuration" and "heteroreduplication" operations on $\mathcal{C}^\star = -\mathcal{C} \cup \mathcal{C}$, respectively, as follows:

$$a \overset{\rho_h}{+} b \;=\; c \Leftrightarrow \eta_b(a) = c,\ \ \eta_b(\rho_h) = b \qquad (2.58)$$
$$\text{and} \quad \eta_b : \mathcal{C}^{\rho_h}_{a} \longrightarrow \mathcal{C}^\star$$

and

$$a \overset{\mu_h}{\times} b = c \Leftrightarrow c \;=\; (a \overset{\rho_h}{+} a \overset{\rho_h}{+} \dots \overset{\rho_h}{+} a)_x : \qquad (2.59)$$
$$x = |b| \ \text{ and } \ f_0 = \rho$$

where $[\rho_h \sim \mu_h] \in \mathcal{I}$, and μ_h is the element in the order of the fundamental interval of the standard nesting level. It follows, from the preceding discussion, that these operations constitute a field with respect to the measurand system $\mathcal{C}^\star$. We call this field the "heterodox field".

If we assume that M is a measurand system and F is an associated fundamental measurand system then if $f(x)$ is a function on M, we say that $b \in R$ is the right-hand (or left-hand) limit of $f(x)$ as x approaches $a \in D$ from the right (or the left) and denote that as $\lim_{x \to a^+} f(x) = b$ (or $\lim_{x \to a^-} f(x) = b$), where R and D are the range and the domain of $f(x)$, respectively. When there exists $\rho \prec \epsilon \in F$

for every $\delta \in F$ such that:

$$a \prec x \prec a \overset{\rho_h}{+} \delta \text{ (or } a \overset{\rho_h}{-} \delta \prec x \prec a) \Rightarrow |f(x) \overset{\rho_h}{-} b| \prec \epsilon \quad (2.60)$$

where the operation "$\overset{\rho_h}{-}$" is defined as follows:

$$a \overset{\rho_h}{-} b = c \Leftrightarrow c \overset{\rho_h}{+} b = a \quad (2.61)$$

the function $f(x)$ is said to have a limit at a if the right-hand and the left-hand limits at a exist and have the same value; that is, if the following condition is satisfied:

$$|x \overset{\rho_h}{-} a| \prec \delta \Rightarrow |f(x) \overset{\rho_h}{-} b| \prec \epsilon \quad (2.62)$$

the function $f(x)$ is called continuous at a given measurand a if it has a limit at a that is equal to the value of $f(x)$ at the same measurand and it is called continuous on an interval if it is continuous at every measurand in that interval (notice that this definition of continuity is consistent with the definition provided above). If $f(x)$ and $g(x)$ are two functions on a measurand system M such that both of them have a limit at the measurand a, then we have:

$$\begin{aligned}
\lim_{x\to a}[f(x) \overset{\rho_h}{+} g(x)] &= \lim_{x\to a} f(x) \overset{\rho_h}{+} \lim_{x\to a} g(x) \\
\lim_{x\to a}[f(x) \overset{\rho_h}{-} g(x)] &= \lim_{x\to a} f(x) \overset{\rho_h}{-} \lim_{x\to a} g(x) \\
\lim_{x\to a}[f(x) \overset{\mu_h}{\times} g(x)] &= \left[\lim_{x\to a} f(x)\right] \overset{\mu_h}{\times} \left[\lim_{x\to a} g(x)\right] \\
\lim_{x\to a} \frac{f(x)}{g(x)} &= \frac{\lim_{x\to a} f(x)}{\lim_{x\to a} g(x)} \quad \left(\lim_{x\to a} g(x) \neq \rho\right) \\
\lim_{x\to a} \sqrt[n]{f(x)} &= \sqrt[n]{\lim_{x\to a} f(x)}
\end{aligned}$$

where for any given two measurands a and b we have $\frac{a}{b} =$

$a \overset{\mu_h}{\div} b$ and $\sqrt[n]{a} = a^{\frac{\mu}{n}} = b \Rightarrow (b \overset{\mu_h}{\times} b \overset{\mu_h}{\times} b \overset{\mu_h}{\times} \dots \overset{\mu_h}{\times} b)_n = a$. To prove the first equality, we suppose that the following limits exist:

$$\lim_{x \to a} f(x) = b \quad \text{and} \quad \lim_{x \to a} g(x) = c$$

so it follows, from limit definition, that for every $\epsilon \succ \rho$ there is $\delta \succ \rho$ such that:

$$|f(x) \overset{\rho_h}{-} b| \prec \frac{\epsilon}{\mu \overset{\rho_h}{+} \mu} \quad \text{when} \quad \rho \prec |x \overset{\rho_h}{-} a| \prec \delta_1$$

and

$$|g(x) \overset{\rho_h}{-} c| \prec \frac{\epsilon}{\mu \overset{\rho_h}{+} \mu} \quad \text{when} \quad \rho \prec |x \overset{\rho_h}{-} a| \prec \delta_2$$

so if we take $\delta = \min(\delta_1, \delta_2)$ then both definitions are satisfied when $\rho \prec |x \overset{\rho_h}{-} a| \prec \delta$ and we have:

$$\begin{aligned} |[f(x) \overset{\rho_h}{+} g(x)] \overset{\rho_h}{-} (b \overset{\rho_h}{+} c)| &= |[f(x) \overset{\rho_h}{-} b] \overset{\rho_h}{+} [g(x) \overset{\rho_h}{-} c]| \\ &\preceq |f(x) \overset{\rho_h}{-} b| \overset{\rho_h}{+} |g(x) \overset{\rho_h}{-} c| \\ &\prec \frac{\epsilon}{\mu \overset{\rho_h}{+} \mu} \overset{\rho_h}{+} \frac{\epsilon}{\mu \overset{\rho_h}{+} \mu} = \epsilon \end{aligned}$$

the previous result is true because the triangular inequality holds for any measurand. The latter fact is a consequence of the definition of the "absolute value" as is shown by the following argument:

$$-|a| \preceq a \preceq |a| \quad \text{and} \quad -|b| \preceq b \preceq |b|$$

and hence

$$-(|a| \overset{\rho_h}{+} |b|) \preceq a \overset{\rho_h}{+} b \preceq (|a| \overset{\rho_h}{+} |b|) \Leftrightarrow |a \overset{\rho_h}{+} b| \preceq |a| \overset{\rho_h}{+} |b|$$

which completes the proof. The second statement follows naturally if we take $-g(x) = h(x)$ while the product statement can be demonstrated by noticing that from the definition of the limits and their existence we have:

$$
\begin{aligned}
|f(x) \overset{\rho_h}{-} b| &\prec \frac{\epsilon}{(\mu \overset{\rho_h}{+} \mu) \overset{\mu_h}{\times} (\mu \overset{\rho_h}{+} |c|)} \\
&= \frac{\epsilon_1}{(\mu \overset{\rho_h}{+} |c|)} \quad \text{when} \quad \rho \prec |x \overset{\rho_h}{-} a| \prec \delta_1 \\
|g(x) \overset{\rho_h}{-} c| &\prec \frac{\epsilon}{(\mu \overset{\rho_h}{+} \mu) \overset{\mu_h}{\times} (\mu \overset{\rho_h}{+} |b|)} \\
&= \frac{\epsilon_1}{(\mu \overset{\rho_h}{+} |b|)} \quad \text{when} \quad \rho \prec |x \overset{\rho_h}{-} a| \prec \delta_2 \\
|g(x) \overset{\rho_h}{-} c| &\prec \mu \ \text{ when } \ \rho \prec |x \overset{\rho_h}{-} a| \prec \delta_3
\end{aligned}
$$

which is true if we take $\delta = \min(\delta_1, \delta_2, \delta_3)$. So the demonstrating argument goes as follows:

$$
\begin{aligned}
|f(x) \overset{\mu_h}{\times} g(x) \overset{\rho_h}{-} b \overset{\mu_h}{\times} c| &= |f(x) \overset{\mu_h}{\times} g(x) \overset{\rho_h}{-} b \overset{\mu_h}{\times} g(x) \\
&\qquad \overset{\rho_h}{+} b \overset{\mu_h}{\times} g(x) \overset{\rho_h}{-} b \overset{\mu_h}{\times} c| \\
&\preceq |f(x) \overset{\mu_h}{\times} g(x) \overset{\rho_h}{-} b \overset{\mu_h}{\times} g(x)| \\
&\qquad \overset{\rho_h}{+} |b \overset{\mu_h}{\times} g(x) \overset{\rho_h}{-} b \overset{\mu_h}{\times} c| \\
&= |g(x)| \overset{\mu_h}{\times} |f(x) \overset{\rho_h}{-} b| \\
&\qquad \overset{\rho_h}{+} |b| \overset{\mu_h}{\times} |g(x) \overset{\rho_h}{-} c| \\
&\prec (\mu \overset{\rho_h}{+} |c|) \overset{\mu_h}{\times} \frac{\epsilon_1}{(\mu \overset{\rho_h}{+} |c|)} \\
&\qquad \overset{\rho_h}{+} (\mu \overset{\rho_h}{+} |b|) \overset{\mu_h}{\times} \frac{\epsilon_1}{(\mu \overset{\rho_h}{+} |b|)} \\
&= \epsilon
\end{aligned}
$$

the quotient statement is a special case of the product statement and, for the last proposition, we have from the product rule:

$$\lim_{x\to a} f(x) = \lim_{x\to a}\left[\sqrt[n]{f(x)}\right]^n = \left[\lim_{x\to a}\sqrt[n]{f(x)}\right]^n$$

so by taking the n_{th} root for both sides, the proposition is established. The function $f(x)$ is said to be "differentiable" if the following limit exists:

$$\lim_{h\to\rho} \frac{f(x \overset{\rho_h}{-} h) \overset{\rho_h}{-} f(x)}{h} = f'(x) = D_x f(x) \tag{2.63}$$

for any differentiable functions $f(x)$ and $g(x)$ on an interval $S \in M$ (where M is a measurand system), the product and quotient of these functions are differentiable on S and we have:

$$\begin{aligned} D_x[f(x) \overset{\mu_h}{\times} g(x)] &= f(x) \overset{\mu_h}{\times} D_x g(x) \overset{\rho_h}{+} g(x) \overset{\mu_h}{\times} D_x f(x) \\ D_x\left[\frac{g(x)}{f(x)}\right] &= \frac{f(x) \overset{\mu_h}{\times} D_x g(x) \overset{\rho_h}{-} g(x) \overset{\mu_h}{\times} D_x f(x)}{f(x) \overset{\mu_h}{\times} f(x)} \end{aligned}$$

where the denominator in the quotient rule is not equal to ρ. The product rule can be established as follows:

$$\begin{aligned} & D_x[f(x) \overset{\mu_h}{\times} g(x)] \\ &= \lim_{h\to\rho} \frac{f(x \overset{\rho_h}{+} h) \overset{\mu_h}{\times} g(x \overset{\rho_h}{+} h) \overset{\rho_h}{-} f(x) \overset{\mu_h}{\times} g(x)}{h} \\ &= \lim_{h\to\rho} \frac{f(x \overset{\rho_h}{+} h) \overset{\mu_h}{\times} g(x \overset{\rho_h}{+} h) \overset{\rho_h}{-} g(x \overset{\rho_h}{+} h) \overset{\mu_h}{\times} f(x)}{h} \\ &\quad \overset{\rho_h}{+} \lim_{h\to\rho} \frac{g(x \overset{\rho_h}{+} h) \overset{\mu_h}{\times} f(x) \overset{\rho_h}{-} f(x) \overset{\mu_h}{\times} g(x)}{h} \\ &= \lim_{h\to\rho} \frac{g(x \overset{\rho_h}{+} h) \overset{\mu_h}{\times} [f(x \overset{\rho_h}{+} h) \overset{\rho_h}{-} f(x)]}{h} \end{aligned}$$

$$
\begin{aligned}
&\overset{\rho_h}{+} \lim_{h\to\rho} \frac{f(x) \overset{\mu_h}{\times} [g(x \overset{\rho_h}{+} h) \overset{\rho_h}{-} g(x)]}{h} \\
&= \lim_{h\to\rho} g(x \overset{\rho_h}{+} h) \overset{\mu_h}{\times} \lim_{h\to\rho} \frac{[f(x \overset{\rho_h}{+} h) \overset{\rho_h}{-} f(x)]}{h} \\
&\overset{\rho_h}{+} \lim_{h\to\rho} f(x) \overset{\mu_h}{\times} \lim_{h\to\rho} \frac{[g(x \overset{\rho_h}{+} h) \overset{\rho_h}{-} g(x)]}{h} \\
&= g(x) \overset{\mu_h}{\times} D_x f(x) \overset{\rho_h}{+} f(x) \overset{\mu_h}{\times} D_x g(x)
\end{aligned}
$$

which is the required result. To deduce the quotient rule, we notice that:

$$
\begin{aligned}
D_x \frac{\mu}{f(x)} &= \lim_{h\to\rho} \frac{\frac{\mu}{f(x \overset{\rho_h}{+} h)} \overset{\rho_h}{-} \frac{\mu}{f(x)}}{h} \\
&= \lim_{h\to\rho} \frac{f(x) \overset{\rho_h}{-} f(x \overset{\rho_h}{+} h)}{f(x) \overset{\mu_h}{\times} f(x \overset{\rho_h}{+} h) \overset{\mu_h}{\times} h} \\
&= \frac{-\mu}{f(x)} \overset{\mu_h}{\times} \lim_{h\to\rho} \frac{\mu}{f(x \overset{\rho_h}{+} h)} \\
&\qquad \overset{\mu_h}{\times} \lim_{h\to\rho} \frac{f(x \overset{\rho_h}{+} h) \overset{\rho_h}{-} f(x)}{h} \\
&= \frac{-D_x f(x)}{f(x) \overset{\mu_h}{\times} f(x)}
\end{aligned}
$$

therefore, using this result and the product rule we have:

$$
\begin{aligned}
D_x \left[\frac{g(x)}{f(x)}\right] &= D_x \left[g(x) \overset{\mu_h}{\times} \frac{\mu}{f(x)}\right] \\
&= g(x) \overset{\mu_h}{\times} D_x \frac{\mu}{f(x)} \overset{\rho_h}{+} \frac{\mu}{f(x)} \overset{\mu_h}{\times} D_x g(x) \\
&= g(x) \overset{\mu_h}{\times} \frac{-D_x f(x)}{f(x) \overset{\mu_h}{\times} f(x)} \overset{\rho_h}{+} \frac{\mu}{f(x)} \overset{\mu_h}{\times} D_x g(x)
\end{aligned}
$$

$$= \frac{f(x) \overset{\mu_h}{\times} D_x g(x) \overset{\rho_h}{-} g(x) \overset{\mu_h}{\times} D_x f(x)}{f(x) \overset{\mu_h}{\times} f(x)}$$

which completes the deduction of the quotient rule. Suppose $F(x) = f[g(x)]$, if $g(x)$ is differentiable on the interval S_1 and $f(u)$ is differentiable on $S_2 = \{u|\ u = g(x)\}$ then $F(x)$ is differentiable on S_1 and we have:

$$D_x F(x) = D_x f[g(x)] = f'[g(x)] \overset{\mu_h}{\times} D_x g(x) \qquad (2.64)$$

we call this statement the "chain rule". To establish this result, we have from the definitions that:

$$\begin{aligned}
D_x F(x) &= \lim_{h \to \rho} \frac{F(x \overset{\rho_h}{+} h) \overset{\rho_h}{-} F(x)}{h} \\
&= \lim_{\Delta x \to \rho} \frac{f[g(x_1)] \overset{\rho_h}{-} f[g(x_2)]}{x_1 \overset{\rho_h}{-} x_2} \\
&= \lim_{\Delta x \to \rho} \frac{f[g(x_1)] \overset{\rho_h}{-} f[g(x_2)]}{x_1 \overset{\rho_h}{-} x_2} \overset{\mu_h}{\times} \frac{g(x_1) \overset{\rho_h}{-} g(x_2)}{g(x_1) \overset{\rho_h}{-} g(x_2)} \\
&= \lim_{\Delta x \to \rho} \frac{f[g(x_1)] \overset{\rho_h}{-} f[g(x_2)]}{g(x_1) \overset{\rho_h}{-} g(x_2)} \overset{\mu_h}{\times} \frac{g(x_1) \overset{\rho_h}{-} g(x_2)}{x_1 \overset{\rho_h}{-} x_2} \\
&= \lim_{\Delta g(x) \to \rho} \frac{f[g(x_1)] \overset{\rho_h}{-} f[g(x_2)]}{g(x_1) \overset{\rho_h}{-} g(x_2)} \\
&\qquad \overset{\mu_h}{\times} \lim_{\Delta x \to \rho} \frac{g(x_1) \overset{\rho_h}{-} g(x_2)}{x_1 \overset{\rho_h}{-} x_2} \\
&= f'[g(x)] \overset{\mu_h}{\times} D_x g(x)
\end{aligned}$$

the previous result holds since $\Delta g(x) \to \rho$ as $\Delta x \to \rho$. If $f(x)$ is a continuous function on the interval $D = [a, b]$ (the notation "$[a, b]$" is used to indicate intervals for "continuous" measurand systems) of the measurand system M, then in the corresponding interval $R = [f(a), f(b)]$, either $f(x) = k$ (where k is a constant and $f(x) \in R$) or $\exists\ x \in D : f(x) \neq k$. So, in both cases, from the continuity of $f(x)$ and the definition of the measurand system, there are $x_1, x_2 \in D$ such that:

$$\begin{aligned} f(x_1) &\succeq f(x) \text{ for any } x \in D \\ f(x_2) &\preceq f(x) \text{ for any } x \in D \end{aligned}$$

from the preceding result, it follows that if $f(x)$ is a continuous function on the closed interval $[a, b]$ (the terminal points are included) and differentiable on the open interval (a, b) (the terminal points are excluded) and $f(a) = f(b) = \rho$, then there is $c \in (a, b)$ such that $f'(c) = \rho$. This proposition can be established if we notice that if $f(x)$ is constant then the statement follows trivially. However, if $f(x)$ is variable, then there is $x_0 \in (a, b)$ such that $f(x_0) \succ \rho$ or $f(x_0) \prec \rho$. If we assume the former case (without loss of generality), then from the continuity of $f(x)$ and the previous result, there is $c \in [a, b]$ such that $f(c) \succeq f(x)$ for every $x \in [a, b]$ and hence $f(c) \succ \rho$ and we have:

$$\begin{aligned} f'_+(c) &= \lim_{h \to \rho^+} \frac{f(c \overset{\rho_h}{+} h) \overset{\rho_h}{-} f(c)}{h} \preceq \rho \\ f'_-(c) &= \lim_{h \to \rho^-} \frac{f(c \overset{\rho_h}{+} h) \overset{\rho_h}{-} f(c)}{h} \succeq \rho \end{aligned}$$

but since $f(x)$ is differentiable on (a, b) then:

$$f'(c) = f'_-(c) = f'_+(c) = \rho$$

and that completes the proof [see figure 2.4 (a)]. This proposition is known as "Rolle's theorem" and can be generalized to the "mean value theorem" by abandoning the assumption $f(a) = f(b) = \rho$. In this case there will be a measurand $c \in (a, b)$ such that:

$$f'(c) = \frac{f(a) \overset{\rho_h}{-} f(b)}{b \overset{\rho_h}{-} a}$$

this result can be proven by constructing the following function [Figure 2.4 (b)] which satisfies the conditions of Rolle's theorem:

$$g(x) = f(x) \overset{\rho_h}{-} f(a) \overset{\rho_h}{-} \frac{f(b) \overset{\rho_h}{-} f(a)}{b \overset{\rho_h}{-} a} \overset{\mu_h}{\times} (x \overset{\rho_h}{-} a)$$

and hence

$$g'(c) = \rho \;\Rightarrow\; f'(c) \overset{\rho_h}{-} \frac{f(b) \overset{\rho_h}{-} f(a)}{b \overset{\rho_h}{-} a} = \rho$$

$$\Rightarrow\; f'(c) = \frac{f(a) \overset{\rho_h}{-} f(b)}{b \overset{\rho_h}{-} a}$$

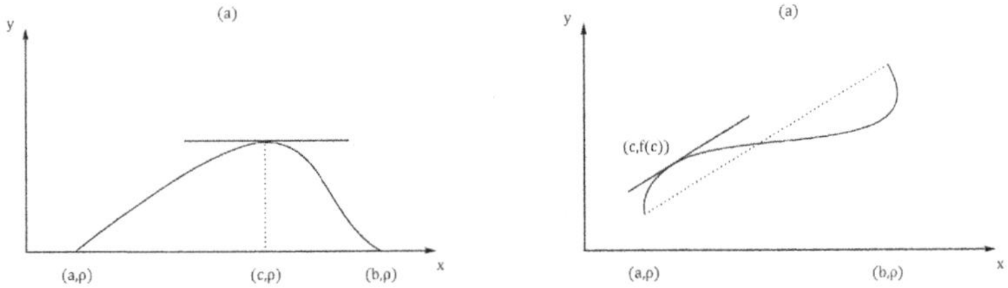

Figure 2.4: (a) The function that satisfies the conditions of Rolle's theorem. (b) The function constructed in the proof of the mean value theorem.

If $f(x)$ is a continuous and differentiable function on the interval $[a, b]$ of any measurand system M, and if we divide that interval into n sub-intervals (Figure 2.5), then for every

$x_i^* \in [x_i, x_{i+1}]$, the Riemann integral is defined as the limits of the following sum:

$$\int_a^b f(x)dx = \lim_{n\to\infty} \sum_{i=1}^{n} f(x_i^*) \overset{\mu_h}{\times} \Delta x \qquad (2.65)$$

where the summation is performed using the operation "$\overset{\rho_h}{+}$". However, since the function $f(x)$ satisfies the mean value theorem conditions on each sub-interval $[x_i, x_{i+1}]$, so if we let $F'(x) = f(x)$, then there will be $c_i \in [x_i, x_{i+1}]$ such that:

$$F(x_i) \overset{\rho_h}{-} F(x_{i-1}) = f(c_i) \overset{\mu_h}{\times} (x_i \overset{\rho_h}{-} x_{i-1}) = f(c_i) \overset{\mu_h}{\times} \Delta x_i$$

and hence

$$\begin{aligned}
\sum_{i=1}^{n} f(c_i) \overset{\mu_h}{\times} \Delta x_i &= \sum_{i=1}^{n} \left[F(x_i) \overset{\rho_h}{-} F(x_{i-1})\right] \\
&= \left[F(x_2) \overset{\rho_h}{-} F(a)\right] \\
&\quad \overset{\rho_h}{+} \left[F(x_3) \overset{\rho_h}{-} F(x_2)\right] \overset{\rho_h}{+} \ldots \\
&\quad \overset{\rho_h}{+} \left[F(x_k) \overset{\rho_h}{-} F(x_{k-1})\right] \overset{\rho_h}{+} \ldots \\
&\quad \overset{\rho_h}{+} \left[F(b) \overset{\rho_h}{-} F(x_{n-1})\right] \\
&= F(b) \overset{\rho_h}{-} F(a)
\end{aligned}$$

and so we have

$$\int_a^b f(x)dx = F(b) \overset{\rho_h}{-} F(a) \qquad (2.66)$$

this result is known as the "fundamental theorem of calculus". The "Riemann-Stieltjes integral" (as a generalization of Riemann integral) and the "Lebesgue integral" along with its generalization (Lebesgue-Stieltjes integral) can be defined for any measurand system using procedures similar to those utilized above. Also all the related theorems can be extended to any measurand system. The existence of these procedures stems from the fact that the preceding concepts were deduced for numbers based on their nature as measurands not based on their nature as numbers.

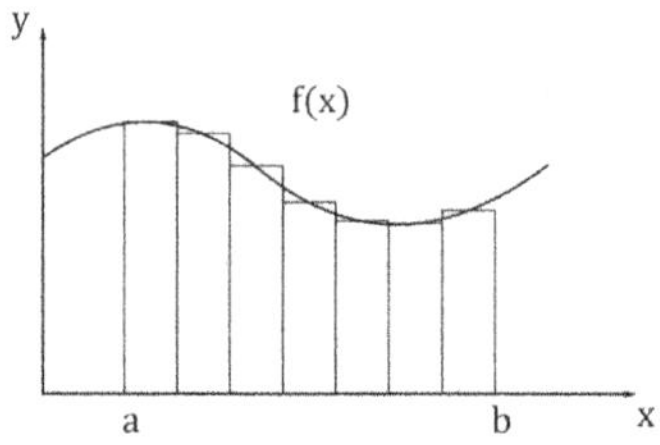

Figure 2.5: The partition of the interval $[a, b]$ into n sub-intervals to form the summation in the definition of Riemann integral.

If A is an ordered set of any concepts (a measurand system) with order n and $\star$ is an operation defined on it which constitutes, with another operation $\circ$, a field on A, then the field $(A, \star, \circ)$ is a redenomination or rearrangement of a heterodox field of the same order created by a heteromensuration and heteroreduplication defined on $B = A$. To establish this proposition, suppose that $f : A \longrightarrow B$ is an isomorphism from A to B (which should exist as we have demonstrated previously) hence we get:

$$f(a \star b) = f(a) \overset{\rho_h}{+} f(b) \text{ and } f(a \circ b) = f(a) \overset{\mu_h}{\times} f(b)$$

so if we change the name of each element using the bijection $x \overset{g}{\mapsto} f(x)$, then it follows that:

$$f(a \star b) \overset{g}{\mapsto} f(a) \star f(b) \text{ and } f(a \circ b) \overset{g}{\mapsto} f(a) \circ f(b)$$

and hence

$$f(a) \star f(b) = f(a) \overset{\rho_h}{+} f(b) \quad \text{and} \quad f(a) \circ f(b) = f(a) \overset{\mu_h}{\times} f(b)$$

therefore, the operations $\star$ and $\circ$ are just other names of the operations $\overset{\rho_h}{+}$ and $\overset{\mu_h}{\times}$ which completes the proof. To illustrate this behavior, let's take the operations $\boxplus$ and $\boxtimes$ which are defined as below:

$$a \boxplus b = a \times b \quad \text{and} \quad a \boxtimes b = a^{\ln b}$$

these operations form a field with respect to the set of positive real numbers $\mathbb{R}^+$. However, if we change the names of the elements in this field by the bijection $x \overset{g}{\mapsto} \ln x$, we will get:

$$a \boxplus b = c = a \times b \Rightarrow \ln c = \ln a + \ln b \Rightarrow c = a + b$$

and

$$a \boxtimes b = c = a^{\ln b} \Rightarrow \ln c = \ln a \times \ln b \Rightarrow c = a \times b$$

hence the operations $\boxplus$ and $\boxtimes$ are just other names of the ordinary addition $+$ and multiplication $\times$, so the field $(\mathbb{R}^+, \boxplus, \boxtimes)$ is a redenomination of the common field of real numbers $(\mathbb{R}, +, \times)$ which constitutes an instance of the orthodox field. However, the latter is merely a special case of the more general concept of the heterodox field. If a measurand system A (an ordered set) with order n constitutes a field with a given operations $\star$ and $\bullet$ then, as we have established, for any other measurand system B with the same order n and any bijection $f : A \longrightarrow B$, there exist two corresponding operations $*$ and $\circ$ that constitute a field on B such that $A \cong B$. Therefore, the operations $*$ and $\circ$ are names (intermediate aspects) of the operations $\star$ and $\circ$ and each element $a \in A$ is a name (intermediate aspect) of the element $f(a) = b \in B$

and hence the field $(B, *, \circ)$ is a name of the field $(A, \star, \bullet)$. The same reasoning can be applied to any ring and even any system of two operations. Furthermore, as we have mentioned, any type of systems can be used as a name for any other type even if they have different orders. However, the most suitable and profitable names of a system are other systems of the same order and type. Hence, naturally, we call the latter type of names the orthodox names. The number of all possible bijections between two measurand systems A and B with the same order n is $n!$. Hence there are $n!$ possible orthodox names (or in short names) for either of them assigned by those bijections. But since the number of possible measurand systems (ordered by nature or by definition) that have the order n is infinite, then the number of possible names for a given measurand system is infinite which is expected since names are intermediate aspects. Any name of these names can be used to describe the others. For example, if the fields $(B, \star, \bullet)$ and $(C, *, \circ)$ are two names of the field $(A, \boxplus, \boxtimes)$, then we can employ the latter to define those names as follows:

$$a \star b = f(f(a)^{-1}) \boxplus f(f(b)^{-1}) \text{ and } a \bullet b = f(f(a)^{-1}) \boxtimes f(f(b)^{-1})$$

and

$$a * b = g(g(a)^{-1}) \boxplus g(g(b)^{-1}) \text{ and } a \circ b = g(g(a)^{-1}) \boxtimes g(g(b)^{-1})$$

where $f : A \longrightarrow B$ and $g : A \longrightarrow C$ are the related two bijections. However, if $\mathfrak{D}$ is the relation of the "definability" of a given name A by another name B in the set of all possible (orthodox) names $\mathcal{N}$ of any field then, according to the preceding discussion, we have:

$$A \mathfrak{D} A \text{ and } A \mathfrak{D} B \Rightarrow B \mathfrak{D} A$$

and if C is a third name we get

$$A\mathfrak{D}B \wedge B\mathfrak{D}C \Rightarrow A\mathfrak{D}C$$

so the definability relation is an equivalence relation that divides $\mathcal{N}$ into equivalence classes and each element of any class of them can be the representative (the definer) of that class. However, the orthodox field follows the natural succession of the elements in any measurand system. Hence it is more convenient to describe any name in a given "nominal" class by its orthodox field. (This selection is a matter of convention and its virtue is a relativistic phenomenon that depends on the constitutional structure of the observer.) As a consequence of this behavior, any name has a measurand structure and algorithms to perform its operations which constitutes, collectively, an arithmetic of that name. As an example, we may take the field $(\mathbb{R}, \oplus, \odot)$ (which is not an orthodox name of the field of real numbers $\mathbb{R}$). The small chart of the operation $\oplus$ is shown in table 2.3 and also the small chart of the operation $\odot$ is provided below (see table 2.4).

$\oplus$	0	0.5	1	2	3	4	5	6	7	8	9
0	−0.5	0	0.5	1.5	2.5	3.5	4.5	5.5	6.5	7.5	8.5
0.5	0	0.5	1	2	3	4	5	6	7	8	9
1	0.5	1	1.5	2.5	3.5	4.5	5.5	6.5	7.5	8.5	9.5
2	1.5	2	2.5	3.5	4.5	5.5	6.5	7.5	8.5	9.5	10.5
3	2.5	3	3.5	4.5	5.5	6.5	7.5	8.5	9.5	10.5	11.5
4	3.5	4	4.5	5.5	6.5	7.5	8.5	9.5	10.5	11.5	12.5
5	4.5	5	5.5	6.5	7.5	8.5	9.5	10.5	11.5	12.5	13.5
6	5.5	6	6.5	7.5	8.5	9.5	10.5	11.5	12.5	13.5	14.5
7	6.5	7	7.5	8.5	9.5	10.5	11.5	12.5	13.5	14.5	15.5
8	7.5	8	8.5	9.5	10.5	11.5	12.5	13.5	14.5	15.5	16.5
9	8.5	9	9.5	10.5	11.5	12.5	13.5	14.5	15.5	16.5	17.5

Table 2.3: The small table of the operation $\oplus$ defined on the set of real numbers $\mathbb{R}$.

These charts can be utilized to deduce the associated algorithms of the operations $\oplus$ and $\odot$ and hence to construct the arithmetic of the field $(\mathbb{R}, \oplus, \odot)$ which then can be used with their isomorphism $(f^{-1} : \mathbb{R} \longrightarrow \mathbb{R})$ to define the ordinary operations of addition and multiplication that constitute with $\mathbb{R}$ the common real field. The constructed arithmetic and the isomorphism between those two fields would produce the following definition:

$$a+b = a \oplus b \oplus 1 \quad \text{and} \quad a \times b = (a \odot b \oplus a \oplus b \oplus 0) \oslash 1.5 \quad (2.67)$$

where the operation $\oslash$ is the inverse of the operation $\odot$. However, as we have mentioned, the best representative of the nominal class is the orthodox name and so, following the conventional practice, we use the common real field $(\mathbb{R}, +, \times)$ to describe the unorthodox field $(\mathbb{R}, \oplus, \odot)$ defined on the real set. Based on the orthodox arithmetic of the ordinary field of real numbers and the isomorphism $(f : \mathbb{R} \longrightarrow \mathbb{R})$ between them , the field $(\mathbb{R}, \oplus, \odot)$ can be defined as follows:

$$a \oplus b = a+b-0.5 \quad \text{and} \quad a \odot b = a \times b + a(a-1) \times (b-1) \quad (2.68)$$

$\odot$	0	1	2	3	4	5	6	7	8	9
0	0	0	−1	−2	−3	−4	−5	−6	−7	−8
1	0	1	2	3	4	5	6	7	8	9
2	−1	2	5	8	11	14	17	20	23	26
3	−2	3	8	13	18	23	28	33	38	43
4	−3	4	11	18	25	32	39	46	53	60
5	−4	5	14	23	32	41	50	59	68	77
6	−5	6	17	28	39	50	61	72	83	94
7	−6	7	20	33	46	59	72	85	98	111
8	−7	8	23	38	53	68	83	98	113	128
9	−8	9	26	43	60	77	94	111	128	145

Table 2.4: The small table of the operation $\odot$ defined on the set $\mathbb{R}$.

As illustrated in the previous example, any name (orthodox or unorthodox field) in a nominal class has a collection of algorithms which constitute its arithmetic and the latter, along with the related isomorphism, can be used to define any other name in that nominal class. However, the standard name (by convention), used to accomplish this task, is the orthodox name and its arithmetic.

As previously stated, in principle, any measurand system can be used to represent any other measurand system. However, the most convenient representation occurs when the two systems have the same order. We call this representation the "orthodox representation" (or in short the representation) of the measurand systems and define it as follows:

$$\forall\, A, B \in \mathcal{M} : A\mathcal{R}B \Leftrightarrow |A| = |B| \qquad (2.69)$$

where $\mathcal{R}$ is the representation relation defined on the set of all measurand systems $\mathcal{M}$. Therefore, according to this definition, for every A, B and $C \in \mathcal{M}$, we have:

$$A\mathcal{R}A, \quad A\mathcal{R}B \Rightarrow B\mathcal{R}A \;\text{ and }\; A\mathcal{R}B \wedge B\mathcal{R}C \Rightarrow A\mathcal{R}C$$

and so the representation relation is an equivalence relation that partitions the set of all measurand systems $\mathcal{M}$ into equivalence classes. We call each equivalence class $\bar{M} \in \mathcal{M}$ a "measurand class" and hence any measurand system in a measurand class can be used to represent that class.

Therefore, the relation that a given name in a nominal class is of the same type constitutes also an equivalence relation on that nominal class. It follows that the orthodox fields cluster into an equivalence class in any nominal class to form what we call the "orthodox cluster", and so any element in the orthodox cluster can be used to represent that cluster and hence to represent the related nominal class. However, by convention, the orthodox cluster is represented by numbers (e.g., $\mathbb{N}, \mathbb{Q}, \mathbb{R}$ and $\mathbb{C}$), so the common fields of numbers are used as names of the relevant orthodox fields. For instance, most of geometrical quantities (lengths, areas, volumes, angles, etc.) and physical quantities (forces, masses, temperatures, luminosities, etc.) are elements of the orthodox cluster that is, conventionally, represented by the ordinary field of real numbers $(\mathbb{R}, +, \times)$. So any of these measurand systems has an orthodox field which can be used to represent this cluster. Moreover, any defined infinite measurand system that constitutes with two operations an orthodox field is an element (a name) of this orthodox cluster and hence can be used to represent it. In fact, any finite subset (finite measurand system) of infinite physical quantity (infinite measurand system) is in the same orthodox cluster of a finite defined measurand system with the same order and hence any of them can be used to represent that cluster. The latter phenomenon can be exploited to describe the physical quantities by a defined finite measurand system with an order restricted by the required degree of precision. The names in the orthodox clusters can be rings or any type of systems with two or more operations and hence the previous discussion holds for any type of systems (names).

2.2.5 Unit systems

Any nesting level of a continuous or discrete measrunad system can be used as a "standard nesting level". We call the fundamental interval of the standard nesting level of a given measurand system a "unit" and the set of all units that belong to a collection of measurand systems is called a "unit system". For a given unit system, the units of the measurand systems that are employed to define the other measurand systems, in the same collection, are called the "fundamental units" and the units of the other measurand systems are called the "derived units". For instance, the meter, kilogram and second are fundamental units in the international system of units (SI) while the units of velocity and work are derived units. For any measurand system, we may choose the set of fundamental intervals in a collection of standard nesting levels of different intervals as nesting units and hence these units will constitute a finite nesting measurand system included in the related measurand system. The nesting units of a measurand system can be homogeneous or fractal as in the international system of units (SI) or inhomogeneous as in the British system of units (the imperial system). The closer the collection of the nesting units is to fractal, the more symmetric and simplified the unit system. The unit system in which a group of units is used as the set of the fundamental units is called an "absolute unit system". For example, the CGS system is an absolute unit system that utilizes centimeter, gram and second as the set of its fundamental units. However, practically, it is sometimes convenient to define a unit system such that the natural units of some measurand systems are taken as the fundamental intervals of their standard nesting levels. We call this process "normalization" and the resulting unit system a "natural unit system". For instance, the-Planck unit system is a natural unit system in which the gravitational constant G, the reduced Planck constant $\hbar$, the speed of light c, the Boltzmann constant k_B and the Coulomb constant k_e are taken to be the fundamental intervals of the standard nesting levels of the relating measurand systems. If a given unit system is constructed such that all the natural units of its measurand systems are defined as the fundamental intervals of the related standard nesting levels, it is called a "pure natural unit system".

2.2.6 The constitutional structure

The ultimate aim of the scientific constituents is to rationalize different entities. Languages, definitions, relations, measurand systems and unit systems rationalize thoughts, meanings, mutual interactions, orders and intensity levels, respectively. We call any rationalized entity the "subject" and refer by the word "vicar" to any intermediate aspect (name) of the scientific constituent that is selected as a representative. We define the set S_c of the ordered pairs (v, s) as follows:

$$S_c = \{(v, s) |\ v \mapsto c \text{ and } s \mapsto c\} \tag{2.70}$$

where $v \mapsto c$ and $s \mapsto c$ are, respectively, a vicar and a subject of the constituent c. The set S_c is called the "structure of the constituent c". If a bijection is established between the union of all possible structures of a given constituent c and the points in the Euclidean plane, we call these points, collectively, the "structural space" of that constituent. Any bijection orders the different vicars and subjects in the coordinate axes and hence creates two measurand systems ordered by definition. If S_c is a structure of a given constituent c, then we say S_c is a "continuous structure" if the following condition is satisfied: $\forall\ (v, s), (v', s') \in S_c\ \exists\ (v'', s'') \in S_c :$

$$v \prec v' \ \wedge\ s \prec s' \Rightarrow v \prec v'' \prec v' \ \wedge\ s \prec s'' \prec s' \tag{2.71}$$

if this condition is not satisfied, we call S_c a "discrete structure". Therefore, any possible structure of a given constituent is a discrete or continuous curve, region or a collection of them in its structural space. The length of the interval between two successive vicars or subjects, in a nesting level, is defined as the order of that interval (since they both are measurand systems). The structure of a constituent c is called a "constant structure" if it represents a horizontal line in the structural space. Otherwise, it is designated as a "variable structure" (see Figure 2.6). If S_c^* is a constant structure of a constituent c and S_c is another structure of the same constituent, and if $|S_c^* \cap S_c| \prec |S_c|$, then S_c is a variable structure. To demonstrate this proposition, let (v, s) be a point in $S_c^* \cap S_c$ and (v', s') be another point in S_c such that $(v', s') \notin S_c^* \cap S_c$ (which is always possible since $|S_c^* \cap S_c| \prec |S_c|$). so, since S_c^* is a constant structure, $v \neq v'$ which establishes the preceding result. The constitutional structure of a given science is defined as follows:

$$\mathcal{S} = \bigcup_{i=1}^{5} S_{c_i} \quad \text{and} \quad S_{c_n} = S_{c_{n'}} \Leftrightarrow n = n' \tag{2.72}$$

where S_{c_i} is one of the five constituents of science. For any given set $A \neq \phi$, we can always define a surjective relation $f : A \longrightarrow \mathcal{S}$ between A and the constitutional structure $\mathcal{S}$ and hence, by this relation, A constitutes a science of a particular subject. Therefore, for any set of concepts (even if it consists of just one element), there are collections of curves and regions in the structural spaces of the constituents of science which form a constitutional structure that renders this set a science.

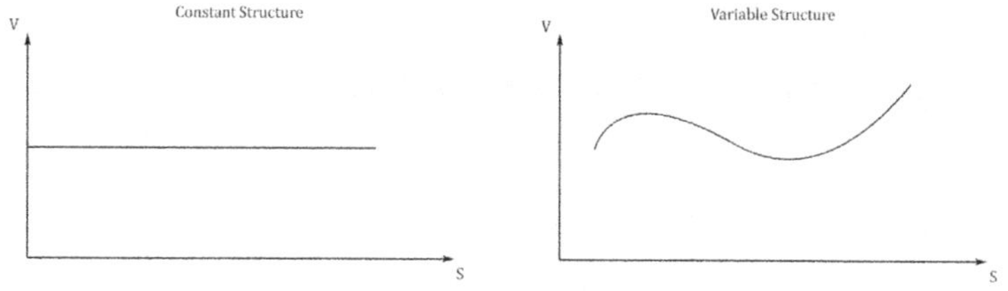

Figure 2.6: constant and variable continuous structures for a given constituent of science.

2.3 The contents of science

Some contents, included in any science, are created by the constituents of that science (hence it is called the essential contents) while others are unitized to create a logical network between its essential contents, to produce strongly supported propositions or to clarify and visualize a line of thought, so they are designated as the accidental contents of science. We will discuss both of these contents in this section.

2.3.1 The essential contents

As previously mentioned, this type of contents arises from the constitutional structure of science and hence they constitute essential components of it. These components are the conceptual vicars, the definitional propositions, the relational vicars, the measurand vicars and the unital vicars. We analyze them in more details below.

Conceptual vicars

The concepts and their sensible connections which create the rational reasoning are captured by language. If we define the "suitability" relation S on the set of all possible languages $\mathfrak{L}$ such that two languages are related by S if both of them are suitable for the constitutional structure of the same observer (orthodox or unorthodox), then for any given three languages a, b and $c \in \mathfrak{L}$, we will have:

$$aSa, \ aSb \Rightarrow bSa \quad \text{and} \quad aSb \wedge bSc \Rightarrow aSc \qquad (2.73)$$

so this relation is an equivalence relation that divides $\mathfrak{L}$ into equivalence classes. We call the representative of any class the vicar of that class. It is obvious that all human languages are included in the equivalence class of the orthodox observer and hence any of them can be used as the vicar of this class. However, the English language is usually utilized as the conceptual vicar (language) of science.

Definitional propositions

If we let $\mathfrak{N}$ to be the set of all possible nomenclatures of all sciences and if we define the nomenclature relation N on that set such that two nomenclatures are related by N if they belong to the same science, then this relation will be an equivalence relation that produces a set of equivalence classes which constitutes a partition of $\mathfrak{N}$. Each of these classes contains all the possible nomenclatures of a particular science. We call any of these classes (denoted by $\bar{N}$) a "nomenclature class" and its representative a "nomenclature vicar". Each nomenclature vicar of a particular science has logical consequences which we call "definitional propositions". For instance, the definition of the area of a region as the number of squares with sides of unit length that cover this region (the ordinary definition; see Figure 2.7) produces, as a logical ramification, the following definitional proposition:

$$\begin{aligned}
A = 2\int_{-r}^{r} \sqrt{r^2 - x^2}\, dx &= 2\int_{-\frac{\pi}{2}}^{\frac{\pi}{2}} r^2 \cos^2\theta \, d\theta \\
&= 2r^2 \int_{-\frac{\pi}{2}}^{\frac{\pi}{2}} \frac{[1+\cos(2\theta)]}{2} \, d\theta \\
&= 2r^2 \left[\frac{\theta}{2} + \frac{\sin(2\theta)}{4}\right]_{-\frac{\pi}{2}}^{\frac{\pi}{2}} \\
&= 2r^2 \left(\frac{\pi}{4} + \frac{\pi}{4}\right) \\
&= \pi r^2
\end{aligned}$$

if we change this definition to describe the area as the number of unit circles that cover the pertinent region, then the area of the circle will be the square of its radius. Both of these definitions are elements of a constituent (a nomenclature) with a constant structure and different vicars. However, if we let the vicar of this nomenclature class varies such that the definition of the area for a given circle with radius r is the number of circles with radius $\sqrt{r}$ which cover the region occupied by that circle, then the area of any circle will be equal to its radius. In fact, for any given number $r \in \mathbb{R}$, there is a curve, a region or a group of them, in the structural space of the nomenclatures, that represents a constant or variable structure, such that the area of any circle is r. These propositions hold for any other regions in the Euclidean space.

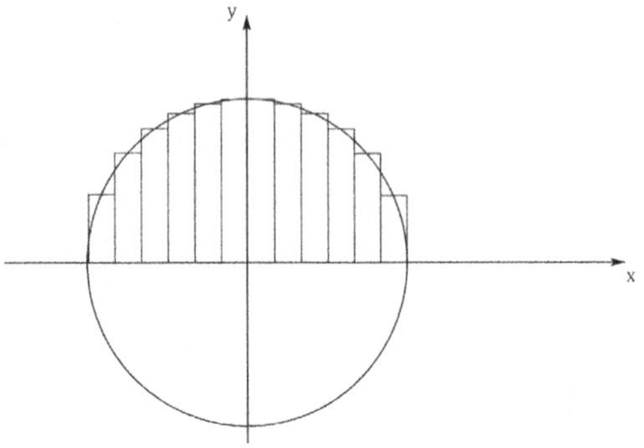

Figure 2.7: The area of the circle as the limit of an approximating sum of areas of circumscribed rectangles.

If we define the relation D on a nomenclature of a given science such that two definitions are related by D if each of them is a definition of the same concept, then this relation will be again an equivalence relation. It follows from this behavior that some definitions form non-trivial clusters (clusters with more than one element) in the nomenclatures of sciences and any element in those clusters can be used to represent its cluster. We call these clusters the "definitional clusters" and their representing elements the "cluster vicars". As an example of the definitional clusters and their vicars, we may take the coordinate systems. A coordinate system is defined as any group of concepts used to describe, quantitatively, the location of a point in a given space with respect to an arbitrary point or group of points. According to this definition, there are an infinitely many possible coordinate systems (one for each given path between these points). However, the nomenclature of many sciences utilizes only four of them (Cartesian coordinate system, polar coordinate system, spherical coordinate system and cylindrical coordinate system; Figure 2.8) and any one of these four coordinate systems is, occasionally, used as a vicar of its cluster depending of the nature of the pertaining problem. When the vicar of this definitional cluster is changed, the group of equations that describe any figure in the related space will have different formations. For example, the equation of the straight line in the two dimensional Euclidean space has the form $y = mx + c$ in the Cartesian coordinate system. However, it will have the form $r,\ \theta = c$ if the polar coordinate system is utilized instead. Another example of the definitions that cluster into one equivalence class is the definitions of the units of a particular quantity. The unit, in this cluster, that is used as the vicar depends on the nature of the problem. As a manifestation of this example, we may take the units of length in the imperial system. The yard, foot and inch are, all of them, units of length, and the choice of any one of these units as the vicar of their cluster is determined by the problem under consideration.

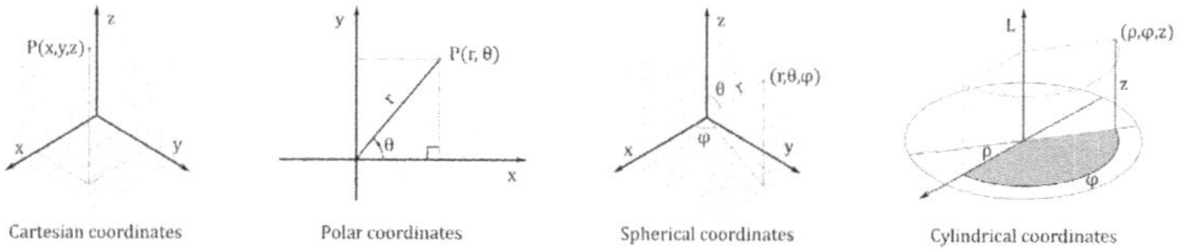

Figure 2.8: The Cartesian, polar, spherical and cylindrical coordinate systems.

The trigonometric functions and their identities are also logical ramifications or definitional propositions of definitions. In the traditional nomenclature of science, these functions are defined as the ratios of the sides of the right triangle which produces the following propositions:

$$\sin\theta = \frac{\text{opposite}}{\text{hypotenuse}}, \quad \cos\theta = \frac{\text{adjacent}}{\text{hypotenuse}} \tag{2.74}$$

and

$$\tan\theta = \frac{\text{opposite}}{\text{adjacent}} \tag{2.75}$$

where the adjacent, opposite and hypotenuse are referred to the side adjacent to θ, the side opposite to θ and the side opposite to the right angle in the same triangle, respectively. These definitions, along with the properties of right triangles, entail all the known trigonometric identities. However, if we define the trigonometric functions using the same definitions, but disregard the restriction of right triangle, so we define the hypotenuse as the side opposite to any angle then, using the sine law, these generalized versions of the trigonometric functions can be deduced as functions of the traditional trigonometric functions as below (see Figure 2.9):

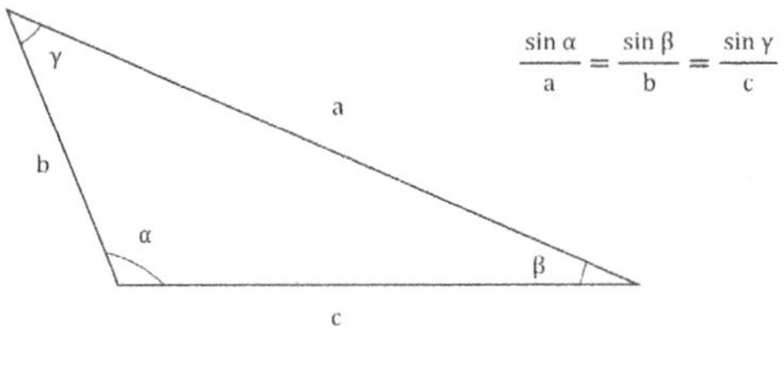

Figure 2.9: A non-right triangle along with the relations between its sides and angles (the law of sines).

$$\sin^{\alpha}\beta = \frac{b}{a} \text{ and } \frac{\sin\beta}{\sin\alpha} = \frac{b}{a} \Rightarrow \sin^{\alpha}\beta = \frac{\sin\beta}{\sin\alpha} \tag{2.76}$$

and

$$\cos^{\alpha}\beta = \frac{c}{a} \text{ and } \frac{\sin\gamma}{\sin\alpha} = \frac{c}{a} \tag{2.77}$$

so

$$\cos^\alpha \beta = \frac{\sin \gamma}{\sin \alpha} = \frac{\sin[180^\circ - (\alpha + \beta)]}{\sin \alpha} = \frac{\sin(\alpha + \beta)}{\sin \alpha} \tag{2.78}$$

and finally we have

$$\tan^\alpha \beta = \frac{\sin^\alpha \beta}{\cos^\alpha \beta} = \frac{\frac{\sin \beta}{\sin \alpha}}{\frac{\sin(\alpha+\beta)}{\sin \alpha}} = \frac{\sin \beta}{\sin(\alpha + \beta)} \tag{2.79}$$

where $\sin(x)$ is taken to be $\sin^{90^\circ}(x)$. This generalized vision of the trigonometric functions assigns different definitions of the same concept, so they form a cluster in a possible nomenclature of mathematics and each of them can be used as the vicar of that cluster. However, the vicar that is utilized in mathematics to represent this cluster is the definition that takes α to be right angle. The latter definitional cluster is inherited by any definitional proposition evolving trigonometric functions. For instance, one important characteristic of light is that it travels between two points along the path that takes the least time. This behavior is called "Fermat's principle". Using this principle, we may deduce a relation between the incident angle β_1 and the refracted angle β_2 in the light refraction phenomenon which is illustrated in Figure 2.10. From this figure we have:

$$|AB| = \sqrt{x^2 + a^2} \quad \text{and} \quad |BC| = \sqrt{(m - x)^2 + b^2}$$

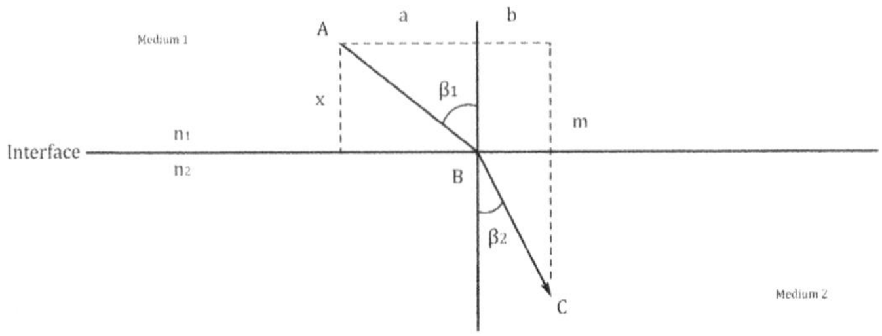

Figure 2.10: Light traveling from medium 1 to medium 2 with different indexes of refraction is refracted at the interface line.

and since the refractive index of a medium is defined as the ratio of the speed of light in vacuum to its speed in that medium, it follows that if we denote the refractive indexes of the first medium and the second medium by n_1 and n_2, respectively, we will have:

$$\begin{aligned} t &= |AB|\times\frac{n_1}{c} + |BC|\times\frac{n_2}{c} \\ &= \sqrt{x^2+a^2}\times\frac{n_1}{c} + \sqrt{(m-x)^2+b^2}\times\frac{n_2}{c} \end{aligned}$$

where t is the time the light takes to travel from A to B. So taking the first derivative of t with respect to x, we get:

$$\begin{aligned} \frac{dt}{dx} &= \frac{n_1 x}{c\sqrt{x^2+a^2}} - \frac{n_2(m-x)}{c\sqrt{(m-x)^2+b^2}} \\ &= \frac{n_1 \sin\beta_1}{c} - \frac{n_2 \sin\beta_2}{c} \end{aligned}$$

and since light obeys Fermat's principle, it follows

$$\frac{n_1 \sin\beta_1}{c} - \frac{n_2 \sin\beta_2}{c} = 0 \Rightarrow n_1 \sin\beta_1 = n_2 \sin\beta_2$$

the latter result, which relates the incident angle and the

refracted angle, is called "Snell's law". However, from the definition of the generalized trigonometric functions we have:

$$
\begin{aligned}
n_1 \sin\beta_1 = n_2 \sin\beta_2 \quad &\Rightarrow \quad n_1 \frac{\sin\beta_1}{\sin\alpha} = n_2 \frac{\sin\beta_2}{\sin\alpha} \\
&\Rightarrow \quad n_1 \sin^\alpha \beta_1 = n_2 \sin^\alpha \beta_2
\end{aligned}
$$

Therefore, this definitional proposition between the incident and refracted angles is a generalized version of Snell's law and it constitutes a definitional cluster for which any given value of α determines one of its elements. However, in the nomenclature of physics, the element of this cluster corresponding to $\alpha = 90°$ is used as the vicar of it.

Relational vicars

As we have established, any operation with number of operands more than two can be reduced to a binary operation, so the latter is the building block of any operation. On the other hand, the equivalence relation O (defined on the set of all operations $\mathfrak{O}$ such that two operations are related by O if they constitute, with respect to a class of representative measurand systems, the same type of operation) divides each of these classes into "operational clusters" (equivalence classes of an equivalence class). The orthodox operations are used to represent any operations and the ordinary operations of numbers are utilized as the vicars of the operational clusters. The measuration in these operations (known as addition) is the base of any other operations, since the multiplication operation (the reduplication operation for numbers) is defined based on addition and all of the other operations are defined based on one of them. However, the number of the possible vicars of any operational cluster is infinity. For instance, the different concentrations of a particular color are ordered by nature and hence they form a measurand system. Any nesting level of this measurand system constitutes, with the mensuration and reduplication operations, a system of two operations, so the orthodox systems constructed with the different nesting levels of this measurand system can be selected as vicars of the related operational clusters. This argument holds for any orthodox system of two operations constructed using any measurand system ordered by nature or by definition. The commonly used vicars of the operational clusters are the real field $(\mathbb{R}, +, \times)$, the complex field $(\mathbb{C}, +, \times)$ and their subfields or any system of a subset of them.

Measurand vicars

We have previously stated that the orthodox representation relation $\mathcal{R}$ divides the set of all measurand systems $\mathcal{M}$ into equivalence classes, the elements of each aggregate into clusters. We, conventionally, use the orthodox cluster to represent each class and a set of ordered numbers to be the vicar of the related orthodox cluster. In principle, there is an orthodox cluster for each class of measurand systems and a numerical representative (vicar) for each orthodox cluster. Therefore, there are numerical measurand systems of all types. For example, the set of natural numbers $\mathbb{N}$ is a discrete infinite measurand system and any subset of it is a discrete finite measurand system, while the set of rational numbers $\mathbb{Q}$ is a continuous infinite measurand system. The different utilized positional numeral systems are examples of fractal measurand systems with infinite frequency (perfect fractal measurand systems). Some numerical systems that are employed to describe particular measurands are inhomogeneous nesting measurand systems with infinite frequency. For instance, the common system used to measure times (years, days, hours, minutes and seconds) is an example of the latter type of measurand systems. However, there are obsolete positional systems that are nonuniform nesting measurand systems with different frequencies. This type of measurand systems change the orthodox distribution pattern of fractionality, so for any integer $n \in \mathbb{Z}$ there are nonuniform nesting measurand systems with frequency m that render n a fraction. To demonstrate this proposition, let the standard nesting level of a numerical measurand system M be the level with frequency m. So if M is a uniform system then any number in that level will be an integer. However, since M is an arbitrary measurand system, it is always possible to choose a nonuniform M such that a given element n in the standard level is a fractional reduplication of the unit (the order of the fundamental interval). Hence the nonuniform system M renders the integer number n (with respect to a uniform system) a fraction and if we change the standard nesting level to the level with frequency $m - 1$, the same argument will be true for the previous unit (the unit of the standard level m) which is a restatement of the proposition set to be proved.

Unital vicars

Let $\mathcal{U}$ be the set of all units. We define the equivalence relation U on that set so that two units are related by U if they are belong to the same messurand system. Therefore, if $p(\bar{u}_i)$ is the power set of the equivalence class (the unital class) $\bar{u}_i$ and $a_i \in p(\bar{u}_i)$, then the previously stated definition of any unit system A is equivalent to the following proposition:

$$A = \bigcup_{i=1}^{n} a_i \tag{2.80}$$

where n is the number of measurands in that unit system (using the set of natural numbers $\mathbb{N}$ as the fundamental measurand system). It follows, as a traditional practice in science, that several collections of vicars are used for each unital class and each of these collections is, conventionally, associated with each other to produce the different unit systems.

2.3.2 The accidental contents

Some contents of science are not produced by its constituents. However, these contents create logical connections between its concepts and propositions and produce highly plausible statements. Also some of them are utilized as tools to visualize and approach some ideas. We call these contents the "accidental contents" of science.

Arguments

The language is a human interface of the nervous signals network created by brain cells which is developed as a response to the observer surrounding conditions, so it is the depository of the rational reasoning. In any language several syllables create a word and the group of words are connected according to specific rules (known as the "syntax") that preserve the rational relations imposed by the developing environment. If these groups of words create a complete meaning that can stand alone, they are called "sentences"; otherwise, they are referred to as "phrases". Some sentences convey assertions such as the sentence "it is now raining " while others just describe meanings that do not indicate any assertions. For instance, the group of words "the sky is beautiful" is a sentence since it creates a complete meaning, but it just describe the feeling of someone without any assertion. The assertive sentences state facts about someone or something and hence these facts may exist or not. We describe the sentence as a "true assertion" if the facts, conveyed by that sentence, do exist. Otherwise, it is designated as a "false assertion". For example, the assertion "the day sky is blue" is a true assertion, and the assertion "the milk is red" is a false one. We call any assertive sentence a "proposition" or a "statement" and its associative status (true or false) the value of that proposition. The most simple statement that any part of which does not constitute an assertion is called a "unit statement". The unit statements can be connected by several connectors to form a "compound statement" which again is an assertion. The connectors that produce compound statements whose values depends only on the values of their unit statements are called the "logical connectors" and the produced compounds are called "logical compound statements". (For brevity, we will refer to them simply as compound statements if there is no confusion.) As an example, the statement "the sky is blue and the night is dark" is a compound statement connected by the logical connecter "and", while the statement "John did not travel to New York because he was ill" is a compound statement that is not connected by a logical connector since its value depends, also, on the causal relation between its unit statements. The conjunction, disjunction, conditional and biconditional are the four basic logical connectors in any language. They constitute equivalence classes which can be represented by many words, but the usual vicars of these classes are "and", "or", "if...then" and "if and only if", respectively. We will utilize the respective symbols $\wedge, \vee, \rightarrow$ and $\leftrightarrow$ to refer to them. For instance, the compound statement "Steve went to the movie theater and George is sleeping" is written symbolically as "$P \wedge Q$", where P stands for the first unit statement and Q stands for the second. The negation of a proposition is denoted by "$\sim$", so if P is the statement "John is walking " then $\sim P$ is its negation "John is not walking". The last two logical connectors ($\rightarrow$ and $\leftrightarrow$) can be defined using the conjunction connecter and the concept of negation as follows:

$$P \rightarrow Q \equiv \sim (P \wedge \sim Q) \tag{2.81}$$

and

$$P \leftrightarrow Q \equiv (P \rightarrow Q) \wedge (Q \rightarrow P) \tag{2.82}$$

the conditional connector which relates a set of statements S_1 that are connected by conjunctions with another set of statements S_2 connected by disjunctions is called a "sequent", so the sequent has the following form:

$$P_1 \wedge P_2 \wedge ... \wedge P_n \to Q_1 \vee Q_2 \vee ... \vee Q_m \tag{2.83}$$

the set S_1 and the set S_2 are called the "antecedents" and the "consequents", respectively. We refer by the word "argument" to any collection of true statements in which a group of propositions (premises) entail or strongly support the truth of a particular statement (called the conclusion), so the argument is a special case of the sequent (a sequent with a single consequent). If we define the relation R on the set of all statements S, such that two elements of S are related by R if they have the same value, then R will be an equivalence relation that divides S into two equivalence classes each of which is called "propositional class". We will extend this quotient set S/R by introducing an imaginary class of statements with "negative truth value". (By definition, the negation of any elements in the imaginary class is again the same element.) Let $\ast$ and ◆ be two operations defined on the set $V = \{-1, 0, 1\}$ as shown in the operational tables below:

$\ast$	-1	0	1
-1	1	-1	0
0	-1	0	1
1	0	1	1

◆	-1	0	1
-1	0	1	-1
0	1	0	0
1	-1	0	1

Table 2.5: The operational tables of the operations $\ast$ and ◆ defined on the set V.

it is obvious from the definitions of these operations that each of them forms a commutative group on the set V. However, for every a, b and $c \in V$, we also have:

$$c \blacklozenge (a \ast b) = c \blacklozenge a \ast c \blacklozenge b = a \blacklozenge c \ast b \blacklozenge c = (a \ast b) \blacklozenge c$$

and hence the operation $\blacklozenge$ distributes over the operation $\ast$, so the system $(V, \ast, \blacklozenge)$ constitutes a field which we will refer to as the "truth field". Therefore, if we define the logical connectors $\wedge$ and $\vee$ on the quotient set S/R such that the bijection $f : S/R \longrightarrow V$ [defined by $(\bar{-t}, \bar{f}, \bar{t}) \overset{f}{\mapsto} (-1, 0, 1)$, where $\bar{-t}$, $\bar{f}$ and $\bar{t}$ are the classes of negative true, false and true propositions respectively] satisfies the following conditions:

$$f(P \vee Q) = f(P) \ast f(Q) \tag{2.84}$$

and

$$f(P \wedge Q) = f(P) \blacklozenge f(Q) \tag{2.85}$$

then f constitutes an isomorphism from S/R to V and hence any argument A composed of unit statements that have "orthodox values" (0 or 1) is an algebraic expression in the truth field and its conclusion is the solution of the equation $A^* = 1$ in that field (where A^* is the premises of the argument A). For every orthodox values x and $y \in V$ we have:

$$\begin{aligned}
x \ast y = 1 \;\Rightarrow\;& (\sim x \blacklozenge \sim y) \blacklozenge (x \ast y) = (\sim x \blacklozenge \sim y) \blacklozenge 1 \\
\Rightarrow\;& (\sim x \blacklozenge \sim y) \blacklozenge x \ast (\sim x \blacklozenge \sim y) \blacklozenge y \\
& \qquad\qquad\qquad\qquad = \sim x \blacklozenge \sim y \\
\Rightarrow\;& \sim y \blacklozenge (\sim x \blacklozenge x) \ast \sim x \blacklozenge (\sim y \blacklozenge y) \\
& \qquad\qquad\qquad\qquad = \sim x \blacklozenge \sim y \\
\Rightarrow\;& \sim y \blacklozenge 0 \ast \sim x \blacklozenge 0 = \sim x \blacklozenge \sim y \\
\Rightarrow\;& (\sim y \ast \sim x) \blacklozenge 0 = \sim x \blacklozenge \sim y \\
\Rightarrow\;& \sim x \blacklozenge \sim y = 0 \\
\Rightarrow\;& \sim (\sim x \blacklozenge \sim y) = 1
\end{aligned}$$

and

$$\begin{aligned}
x * y = 0 \;\Rightarrow\;& (\sim x \blacklozenge \sim y) \blacklozenge (x * y) = (\sim x \blacklozenge \sim y) \blacklozenge 0 \\
\Rightarrow\;& (\sim x \blacklozenge \sim y) \blacklozenge x * (\sim x \blacklozenge \sim y) \blacklozenge y \\
& \qquad\qquad = \sim x \blacklozenge \sim y \blacklozenge 0 \\
\Rightarrow\;& \sim y \blacklozenge (\sim x \blacklozenge x) * \sim x \blacklozenge (\sim y \blacklozenge y) \\
& \qquad\qquad = \sim x \blacklozenge \sim y \blacklozenge 0 \\
\Rightarrow\;& \sim y \blacklozenge 0 * \sim x \blacklozenge 0 = \sim x \blacklozenge \sim y \blacklozenge 0 \\
\Rightarrow\;& (\sim y * \sim x) \blacklozenge 0 = \sim x \blacklozenge \sim y \blacklozenge 0 \\
\Rightarrow\;& \sim x * \sim y = \sim x \blacklozenge \sim y \\
\Rightarrow\;& \sim x \blacklozenge \sim y = 1 \\
\Rightarrow\;& \sim (\sim x \blacklozenge \sim y) = 0
\end{aligned}$$

and so we have

$$x * y = \sim (\sim x \blacklozenge \sim y) \tag{2.86}$$

Using the same line of thought it can be shown that:

$$x \blacklozenge y = \sim (\sim x * \sim y) \tag{2.87}$$

these two results are called "De Morgan's laws". To illustrate the previous discussion let's take the following argument as an example:

$$(P \wedge Q) \wedge [(P \vee Q) \rightarrow R] \Rightarrow X \tag{2.88}$$

we know, by definition, that

$$(P \vee Q) \rightarrow R \equiv \sim [(P \vee Q) \wedge \sim R] \tag{2.89}$$

so, if we let x, y and z be the images of P, Q and R in the truth field V, the conclusion of this argument will be the solution of the following equation in V:

$$(x \blacklozenge y) \blacklozenge \sim [(x \ast y) \blacklozenge \sim z] = 1$$

therefore we have

$$\begin{aligned} 1 \blacklozenge \sim [(x \ast y) \blacklozenge \sim z] = 1 \;&\Rightarrow\; \sim [(x \ast y) \blacklozenge \sim z] = 1 \\ &\Rightarrow\; (x \ast y) \blacklozenge \sim z = 0 \\ &\Rightarrow\; 1 \blacklozenge \sim z = 0 \\ &\Rightarrow\; \sim z = 0 \\ &\Rightarrow\; z = 1 \Rightarrow X = R \end{aligned}$$

The preceding discussion constitutes the so called propositional logic. The traditional concepts, used to treat arguments in this domain of logic, are the truth tables. These tables contain only the orthodox values of the propositions, so any proposition is either true (denoted by T) or false (denoted by F). Using the previous definition of the logical connectors, the truth table of $\wedge$ and $\vee$ connectors will be:

P	Q	P ∧ Q
T	T	T
T	F	F
F	T	F
F	F	F

P	Q	P ∨ Q
T	T	T
T	F	T
F	T	T
F	F	F

Table 2.6: The truth tables of the logical connectors $\wedge$ and $\vee$.

and hence, by definitions, the truth tables of the conditional $\rightarrow$ and biconditional $\leftrightarrow$ connectors will be:

P	Q	P $\to$ Q
T	T	T
T	F	F
F	T	T
F	F	T

P	Q	P $\leftrightarrow$ Q
T	T	T
T	F	F
F	T	F
F	F	T

Table 2.7: The truth tables of the logical connectors $\to$ and $\leftrightarrow$.

Any compound statement is a combination of these logical connectors and hence it has a truth table that is constructed from the truth tables of its constitutional connectors. A logical expression is called a "tautology" if all of its truth values in its table are Ts. A given argument is defined to be true if its corresponding sequent is a tautology. If we return to the previous argument, we can construct the following truth table:

P	Q	R	P $\wedge$ Q	P $\vee$ Q	(P $\vee$ Q) $\to$ R	A^*	$A^* \to$ R
T	T	T	T	T	T	T	T
T	T	F	T	T	F	F	T
T	F	T	F	T	T	F	T
T	F	F	F	T	F	F	T
F	T	T	F	T	T	F	T
F	T	F	F	T	F	F	T
F	F	T	F	F	T	F	T
F	F	F	F	F	T	F	T

Table 2.8: The truth table of the previous argument $A \equiv A^* \to R$, where A^* is the premises of the argument.

it follows, from the last column of this table, that the argument A is a tautology and hence it is a true argument. The propositional arguments can, also, be treated using the so called " natural deduction". In this method, a set of rules derived from the definitions of the logical connectors is used to deduce intermediate conclusions that are utilized with the premises to establish the argument. These rules can be classified into four groups. The first group includes the hypothesized statements (the premises and assumptions) and the second group pertains to inferences from a single statement, while the third group includes the inferences from several statements and, finally, the last group contains the rules of deduction based on the assumptions. The rules in the first group (rule of premises RP and rule of assumptions RA) are just organizing rules and have no intermediate conclusions that result from their applications. However, the remaining three groups have inferences and we will state them as below:

Group(2)

Simplification (S): $P \wedge Q \Rightarrow P,\ Q$ Addition (A): $P \Rightarrow P \vee Q$

Substitution (Sub): $P,\ P \equiv Q \Rightarrow Q$

Group (3)

Disjunctive syllogism (DS): $(P \vee Q) \wedge \sim P \Rightarrow Q$

Modus ponens (MP): $(P \rightarrow Q) \wedge P \Rightarrow Q$

Modus tollens (MT): $(P \rightarrow Q) \wedge \sim Q \Rightarrow \sim P$

Conjunction (C): $P,\ Q \Rightarrow (P \wedge Q)$

Hypothetical syllogism (HS): $(P \rightarrow Q) \wedge (Q \rightarrow R) \Rightarrow P \rightarrow R$

Constructive dilemma (CD): $(P \rightarrow Q) \wedge (R \rightarrow S) \Rightarrow (P \vee R)$

Group (4)

Reductio (Red): Assuming $P \Rightarrow (P \rightarrow Q) \wedge \sim Q \Rightarrow \sim P$

Conditionalization (Cd): Assuming P and Q based on it $\Rightarrow (P \rightarrow Q)$

these rules along with adequate intellectual skills can be used to establish the validity of any argument. If we apply these rules to the previous argument (argument 2.82), we will get:

1. $P \wedge Q$ RP
2. $(P \vee Q) \rightarrow R$ RP
3. P S_1
4. Q S_1
5. $P \vee Q$ A_3
6. R $\mathrm{MP}_{2,5}$

where RP, S_1, A_3 and $\mathrm{MP}_{2,5}$ stand for the related rules which are the rule of premises, simplification, addition and modus ponens, respectively. This result agrees with those obtained by using the previous two methods. The other logical domain which forms, with the propositional logic, the "deductive logic" is the "predicate logic". Some statements indicate the inclusion or exclusion of all or part of the extension of a particular word from the extension of another word. We call the first extension the "subject" of that statement and the second its "predicate". For instance, in the statements "all birds have wings" and "some animals are not mammals" the words "birds" and "animals" are the subjects of these statements, respectively, and the words "wings" and "mammals" are the respective predicates of them. However, the first statement indicates a total inclusion and the second indicates a partial exclusion. Therefore, the nature of these statements depends on their "quantity" (part or all) and their "quality" (inclusion or exclusion - we refer to the inclusion as the "positive quality" and the exclusion as the "negative quality"), and hence there are the following four possible types of them:

1. All S is included in P

2. All S is excluded from P

3. Some S is included in P

4. Some S is excluded from P

we call these four types A, E, I and O, respectively. An argument in the predicate logic (called a "syllogism") is defined as a set of three true statements of the latter type that contains, collectively, three extensions. This set should satisfy the condition that two of its statements (called the premises) have a common extension, while the other two uncommon extensions of these statements form the subject and predicate of the third statement (called the conclusion). We call the word that has the common extension the "middle term" and the words of the extensions of the subject and predicate of the conclusion are referred to as the minor and major terms, respectively. A term is designated as a distributed term if it is totally included or excluded from another related term. Otherwise, it is referred to as an undistributed term. The "mood" of the syllogism is defined as follows:

$$\mathrm{Mood} = \mathrm{XYZ} : \mathrm{X}, \mathrm{Y}, \mathrm{Z} \in \{\mathrm{A}, \mathrm{E}, \mathrm{I}, \mathrm{O}\} \tag{2.90}$$

according to the definition of the predicate argument (the syllogism), the middle term has four possible positions in the two premises of the argument. The "figure" of a syllogism is defined below as one of these possible positions:

$$\begin{aligned} \mathrm{Figure}(n) &= (\mathrm{X}, \mathrm{Y})_n : (\mathrm{X}, \mathrm{Y})_n \in \{(\mathrm{S}, \mathrm{P})_1, \\ &\qquad (\mathrm{P}, \mathrm{P})_2, (\mathrm{S}, \mathrm{S})_3, (\mathrm{P}, \mathrm{S})_4\} \end{aligned} \tag{2.91}$$

by definition, there are $4 \times 4 \times 4$ moods and 4 figures for each mood. Hence the number of possible predicate arguments for each set of three extensions is 256. The validity of a predicate argument (syllogism) depends on the argument nature and the relations between its three extensions. Therefore it depends on the quantity and the quality of the argument. However, any syllogism that satisfies the following six rules is considered a valid syllogism:

1. The argument must has exactly three terms.
2. If all the premises are positive the conclusion must also be a positive statement.
3. If all the premises are negative the conclusion must be invalid.
4. If only one premise is negative the conclusion must be negative.
5. The distributed term in the conclusion must be distributed in the related premise.
6. The middle term must be distributed in the premises at least one time.

therefore, according to these rules, only 15 out of 256 arguments are valid and if we adopt the assumption that no extension is an empty set (the Aristotelian assumption), this number would increase to 24 arguments. The predicate arguments can, alternatively, be treated by using the concept of "Venn's diagram". In this method, Venn diagram is used as a visual intermediate aspect (name) of the extension of any word or phrase and the universal set of those extensions (which is called the "universe of discourse") is again represented by Venn diagram. Therefore, since any "valid" syllogism has exactly three terms (the major, the minor and the middle terms), each argument divides the universe of discourse into eight regions (see Figure 2.11). The universal and existential quantifiers, along with the propositional logic connectors, can be utilized as a collection of vicars for the linguistic equivalence classes of the relation E defined as follows:

$$aEb : a_m = b_m \quad \text{and} \quad x_m \in \mathcal{E}_l \tag{2.92}$$

where x_m is the meaning of the word or the phrase x and $\mathcal{E}_l$ is the set of meanings of all the words and the phrases in a given language l.

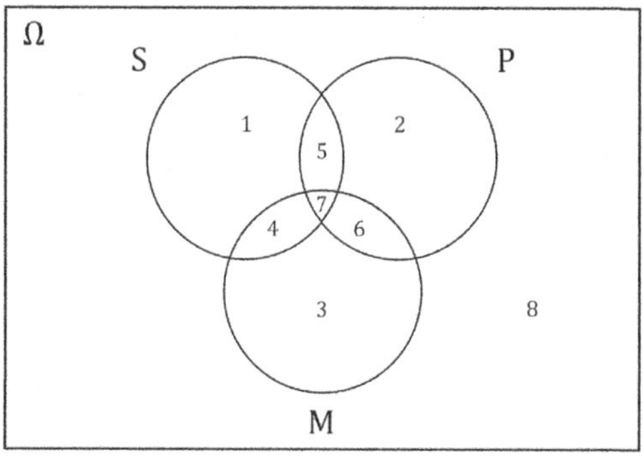

Figure 2.11: The Venn's diagram representation (visual name) of a given argument.

the total positive and negative (inclusion and exclusion) statements are represented, diagrammatically, by shadowed and empty regions, respectively, while the partial positive and negative statements are represented by arrows pointing to the related regions. Using these vicars, any predicate argument can be expressed symbolically. For example, the premises of the syllogism "if all M is P and some S is M then some S is P" can be represented using those vicars as follows:

$$\forall\, x \in \Omega : x \in M \rightarrow x \in P \ \wedge \ \exists\, x \in \Omega : x \in S \rightarrow x \in M$$

and its stated conclusion as

$$\exists\, x \in \Omega : x \in S \rightarrow x \in P$$

the valid predicate argument, in this method, is the one which the name (Venn diagram) of the assertion of its conclusion is included in the name (Venn diagram) of the assertion of all its premises. As an example, from the Venn diagrams of the premises and the conclusion of the previous argument (Figure 2.12), it is clear that the preceding argument is a valid syllogism.

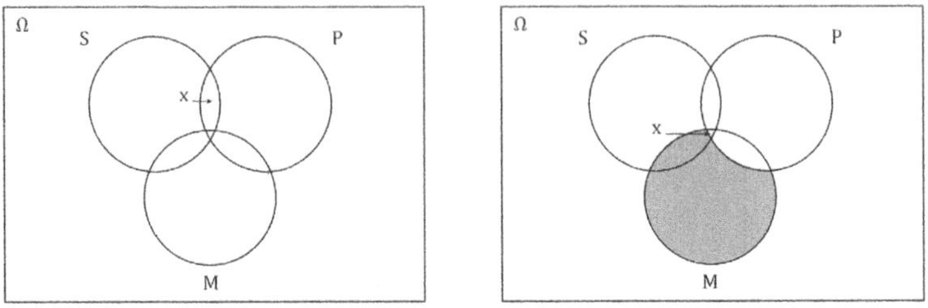

Figure 2.12: The Venn's diagram representations (visual names) of the argument "if all M is P and some S is M then some S is P", where the arrows and the shadows indicate regions with some variables and empty regions, respectively.

The propositional and predicate logic are called collectively (as we have mentioned) the deductive logic. Any argument that entails the truth of a statement as a logical ramification of the truth of a collection of other statements is called a deductive argument. It follows that the deductive logic produces a logical network between the constituents and the essential contents of science.

Non-definitional propositions

Logical reasoning is not restricted to the deductive arguments in which the truth of a group of premises entails the truth of the conclusion since the line of reasoning in which the truth of a group of statements (the premises) "strongly" supports the truth of another statement (the conclusion) is also considered a logical reasoning. This type of logic is called the "inductive logic". Therefore, the difference between deductive arguments and inductive arguments is the degree of uncertainty in the truth of the supported statement (conclusion). While the conclusion in the deductive argument is "absolutely" certain, it has some "uncertainty" in the inductive arguments. We call the conclusion of any inductive argument a "non-definitional proposition". When we encounter a problem, we appeal to a body of knowledge and previous experience pertinent to that problem which suggest a set of solutions to it. We call that body of knowledge and previous experience the "background knowledge" of the problem and any suggested solution a "hypothesis". The set of all plausible solutions suggested by the background knowledge is called the "initially plausible hypotheses". Any conclusion (hypothesis) of an inductive argument is a predicate assertion of the form "n S's are P's", where "n" is a quantifier which is usually a percentage. To confirm one of the initially plausible hypotheses by disconfirming its competitors, we have to collect some data from the extension of the subject of the hypothesized conclusion which we call the "population" of the research. However, in many cases it is impossible to collect data about all of the population of the research since this will destroy the whole population. For example, testing the expected age of a light bulb by examining the whole production (the population) will destroy it. In other cases the population is very large or contains infinitely many members such as the extension of the phrase "adult human" and "the products produced by a company forever". For these reasons and to reduce the high cost evolving some researches, we have to select a sample from the population. But this sample has to be chosen in an appropriate way such that if a characteristic is present or absent in the population, it should be present or absent in the sample. This sample has to be also typical of the population and accurately recorded. We call this type of sample a "representative sample" or a "random sample". To calculate the fraction of S that are Ps is simply to divide the number of elements in the extension of S that are Ps by the total number of the sample. However, since the population of the research has usually a large number of the same element, each of these elements has a number of "frequency" that constitutes, collectively, the "frequency distribution" of the related population. If we normalized this distribution (divide the frequency of each of its elements by the total number of frequencies), the normalized frequency of a given element will coincide with the definition of the probability of that element, and hence the normalized frequency distribution is called the "probability distribution". We call any element of a particular probability distribution the "random variable" and the related probability distribution is called discrete or continuous if this random variable is discrete or continuous, respectively. If there is no reason to distinguish the likelihood of one element of a given population from another then we assume that each element has the same frequency (we call this assumption the "principle of insufficient reason") and the resulting probability distribution is called a "uniform probability distribution". However, if the probability distribution of the random variable of the subject extension in the related hypothesis deviates from "uniformity" (the principle of insufficient reason is violated) then this deviation indicates the existence of a reason that creates the observed distinction (see Figure 2.13). Two entities have causal or statistical connections if the existence of one of them entails the existence of the other. This entailment could exist by definition or by nature. In the first case the connection is regarded as a statistical connection. For instance, there is a strong correlation between the traffic hour in New York city and the tidal effect in some areas in the globe, but this correlation is created by definition (if the time tables of schools and businesses are changed, the connection will vanish) so the connection is statistical. However, in the second case, the persistence of the constancy of the natural entailment that produces the connection varies for different circumstances, and it is also a relativistic phenomenon that depends on the constitutional structure of the observer.

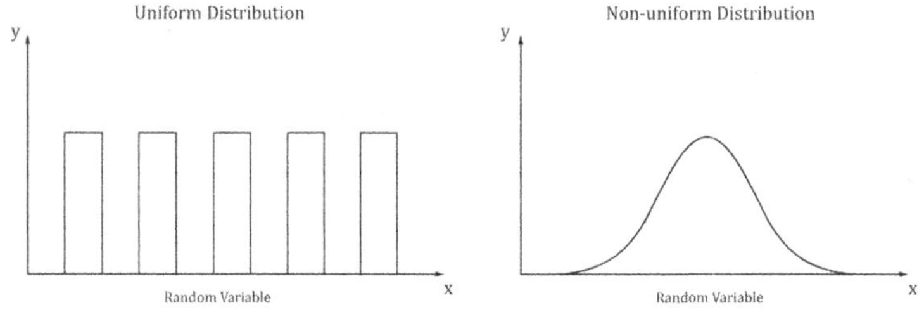

Figure 2.13: A uniform probability distribution and the normal distribution which constitutes an example of the non-uniform probability distribution.

We call the natural entailment which has a constancy that persists for a short time relative to the standard observer (human being as an orthodox observer) a "transit entailment" and that which lasts for a long time relative to the standard observer a "permanent entailment". The connection which results from transit entailment is regarded as a statistical connection, while that which results from a permanent entailment is considered as a causal connection.

Any hypothesis in the set of initially plausible hypotheses has an associative probability distribution and since it is assumed that any initially plausible hypothesis describes a possible set of characteristics of the related population then any random sample drawn from that population should appear as if it was drawn from the population pertinent to the probability distribution associated with the most plausible hypothesis. The hypothesis is disconfirmed if the random sample has a probability less then 5% or 1% according to the probability distribution associated with that hypothesis and it is confirmed if the representative sample has a probability of 95% or 99% in that distribution. The decision about which of these criteria should be applied depends on the discipline on which the research is conducted. In humanities, we utilized the first criterion (5% for disconfirmation and 95% for confirmation) and in natural sciences, we apply the second criterion. The confirmed hypothesis based on the preceding procedures constitutes the conclusion (the non-definitional proposition) of the inductive argument and the background knowledge along with the disconfirming data constitute the premises of that argument. This type of argument is called induction because the conclusion is strongly supported or "justified" by the premises since the set of initially plausible hypotheses does not include all the possible solutions of the problem. However, if it did include all the possible solutions, then the conclusion would be entailed by the premises and the argument would be a deduction whose conclusion is a definitional proposition. The measurand systems utilized to formalize the previous concepts (the probability distributions and the confirmation criteria) are sets of numbers which are vicars of the orthodox clusters. However, any other members of the orthodox clusters or the unorthodox clusters can be used as well to accomplish this task.

Pictures, graphs and diagrams

The visual genuine and intermediate aspects are used extensively in science. They serve as a guide and refinement tools of the thought created by the logical reasoning that connects the essential contents of science. Pictures are photographs or drawings that describe the genuine aspects of the meanings, while the graphs are intermediate aspects (names) of relations described by vicars of the orthodox or unorthodox clusters, and finally the diagrams are visual middle aspects (visual names) of a structure or process. Pictures, graphs and diagrams are called, collectively, "figures".

2.4 Science, Pseudoscience and Fantasy

As we have stated above, science is a group of correlated and isolated concepts pertinent to a set of rational data and arising from the constancy of the environment of the scientific objects. Therefore, the scientific environment is the extension (the reality) of the group of concepts which constitutes a science. If an assumed extension of a given group of concepts does not coincide with the actual reality of that group, and if the inconsistency is subtle, this group is called a "pseudoscience" of that reality. However, if the inconsistency is sharp, that given group of concepts is called a "fantasy". So according to the preceding definitions, any set of concepts is a science, pseudoscience and fantasy depending on the assumed extensions (realities) of these concepts. For instance, physics is a science of the "living" natural phenomena, but it is a pseudoscience relative to a hypothetical universe with nature differing slightly from the nature of our universe, and it is a fantasy with respect to a universe with nature that contrasts the nature of our universe. On the other hand, the set of concepts created by the science fiction and fantasy movies are sciences with respect to the assumed extensions (realities) in those movies, but they are, respectively, pseudosciences and fantasies relative to the living reality.

2.5 Mind, Science and Logic

The system is a set of concepts that interact with each other to serve as or produce a particular concept. For example, the chair is a system since it consists of several concepts (nails, wood, leather, glue, etc.) that interact with each other to serve as a device of sitting. Another example of the system is what we call the "intellectual system" in which mind, science and logic interact with each other to reach the scientific objects. We define "mind" as any concept that has the capability of taking advantage of the science by using logic to reach the scientific objects. Any concept that can accomplish this task is called a mind regardless of whether it is made up of living tissues or electric circuits. If the mind when performing its intellectual process is supported externally or internally by another concept, we call this subsystem an "aided mind"; otherwise, it is referred to as an "unaided mind". It is obvious that for any measurand system and mind there is a supporting concept which constitutes with this mind an aided mind that is suitable for the given measurand system. Therefore, the numerical description of the different quantities is very suitable for the unaided human mind. However, for the aided human mind, any measurand system from the orthodox or heterodox clusters is a suitable vicar to describe the nature.

Epilogue

The nature of existence is a relativistic phenomenon which depends on the constitutional structure of the observer. The abstract treatment of any concept reveals that it also constitutes a reality, science, existence and attribute. The surrounding circumstances in which the observer has developed determine the nature of its senses, and we are an example of the orthodox observer that has been developed in an environment which supplied him with the ability to distinguish between seven types of meanings. The Adamic sciences are intermediate aspects (names) of the extension of the sensibles as a reality. For any given group of concepts, there are more than one scientific constitutional structure that renders this group a science. Any group of concepts is science, pseudoscience and fantasy with respect to the appropriate extensions (realities). Any concept that is capable of employing science by using logic to reach the scientific objectives is a mind that forms with science and logic the intellectual system. In this book we have developed a systematic approach to treat any science as a science. Therefore the virtues of this theory are like the virtues of any science which are: removing the ambiguity of the related concepts, producing an appreciation of their logical connections, exploring the entire reality of the related science and explaining the phenomena pertinent to that science. If we take as an example the equation that describes the centrifugal force applied on an object in a rotational motion as a function of its mass, its velocity and the radius of the rotational orbit, the derivation of this formula is a mathematical phenomenon which is explained by mathematics. The reason why this derivation is true is a logical phenomenon that is investigated in logic. The existence of a particular linguistic compositions and its constituents is a linguistic phenomenon that is explained in languages. The reason, the origin and the development of the meaning of a given word in that derivation is an etymological phenomenon which is described by etymology. And the reason why that particular equation with its numerical and operational constituents stands for the centrifugal force is a scienatic phenomenon that is described by this science.

INDEX

Made in the USA
Monee, IL
04 November 2026

67384801R00075